Wolfgang Hackbusch

Efficient Solutions of Elliptic Systems

Notes on Numerical Fluid Mechanics
Volume 10

Series Editors: Ernst Heinrich Hirschel, München
Maurizio Pandolfi, Torino
Arthur Rizzi, Stockholm
Bernard Roux, Marseille

Manuscripts should have well over 100 pages. As they will be reproduced foto-
mechanically they should be typed with utmost care on special stationary which
will be supplied on request. In print, the size will be reduced linearly to approximately 75 %. Figures and diagrams should be lettered accordingly so as to produce
letters not smaller than 2 mm in print. The same is valid for handwritten formulae.
Manuscripts (in English) or proposals should be sent to the general editor Prof.
Dr. E. H. Hirschel, MBB-LKE 122, Postfach 80 11 60, D–8000 München 80.

Wolfgang Hackbusch (Ed.)

Efficient Solutions of Elliptic Systems

Proceedings of a GAMM-Seminar
Kiel, January 27 to 29, 1984

Springer Fachmedien Wiesbaden GmbH

CIP-Kurztitelaufnahme der Deutschen Bibliothek

Efficient solutions of elliptic systems:
proceedings of a GAMM seminar Kiel,
January 27 to 29, 1984 / Wolfgang Hackbusch
(ed.).
   (Notes on numerical fluid mechanics; Vol. 10)
   ISBN 978-3-528-08084-6   ISBN 978-3-663-14169-3 (eBook)
   DOI 10.1007/978-3-663-14169-3

NE: Hackbusch, Wolfgang [Hrsg.]; Gesellschaft
für Angewandte Mathematik und Mechanik; GT

Produced by Industrie u. Verlagsdruck, Walluf b. Wiesbaden

ISBN 978-3-528-08084-6

FOREWORD

The GAMM-Committee for Efficient Numerical Methods for PDE
(GAMM-Fachausschuß "Effiziente numerische Verfahren für partiel-
le Differentialgleichungen") organises conferences and seminars
on subjects concerning the algorithmic treatment of pde problems.

The topic "EFFICIENT SOLUTION OF ELLIPTIC SYSTEMS" of a GAMM-
Seminar held at the University of Kiel, January 27-29, 1984,
plays a central rôle in structural and fluid mechanics. For instance, such elliptic systems are formed by the equations of
Stokes and Navier-Stokes. The discretisation eq by finite elements and the iterative solution of the arising discrete equations are more difficult than for single elliptic equations.

The Seminar was attended by 43 scientists from 12 countries.
It was possible to bring together experts in particular from
the fields of finite element methods and multi-grid methods.
Accordingly most of the 17 papers presented at the Seminar con-
cerned the finite element discretisation and the multi-grid
solution of the discretised problems. These proceedings contain
11 contributions in alphabetical order.

The editor, who has also been the organiser of this Seminar,
would like to acknowledge a support from the land Schleswig-
Holstein.

Kiel, April 1984   W. Hackbusch

CONTENTS

Page

The Direct Solution of a Generalized Biharmonic Equation on a Disk

Petter Bjørstad

Veritas Research
N-1322 Høvik
NORWAY

ABSTRACT

An efficient, direct solution algorithm for a generalized biharmonic equation on a disk is described. The approximation is second order accurate and the computational work is essentially proportional to the number of grid points. This work is motivated by the usefulness of such a solver in the numerical study of a more complicated model equation describing non-linear pattern formation near the onset of Rayleigh-Benard convection [1].

1. Introduction

In recent years a substantial effort has been devoted to the development of highly effective algorithms for solving certain classes of elliptic partial differential equations. The algorithms known as fast Poisson solvers, started this line of development. Fast Poisson solvers have been implemented into high quality computer software and are now widely available [2]. The class of problems that can be handled have steadily increased. Reliable software based on multigrid techniques [3] have been produced for both rectangular and nonrectangular domains [4], solving general second order elliptic problems, thus extending the smaller class of separable problems originally solved by fast Poisson solvers. The work on capacitance matrix methods [5],[6] is another line of development taking advantage of very special solvers in a more general problem setting. Recent work on algorithms for sub-structured problems [7] represents another approach for new extensions of the available algorithms in this area.

These codes have gained considerable acceptance as algorithmic building blocks when solving more complicated problems. It may well be true that computationally optimal algorithms can be developed for a large number of special problems that are of interest, but often the time and effort to build such codes are large compared with the importance of the problem at hand. It may then be better to make an algorithm based on standard, highly efficient subroutines for subproblems that the given problem can be broken into.

Efficient methods for fourth order elliptic problems, and corresponding computer software, are less widely available. An algorithm with the same favorable operation count for the biharmonic equation in rectangular geometry, has been developed [8]. One application of this algorithm is the study of nonlinear pattern formation near the onset of Rayleigh-Benard convection [9]. This phenomenon can be modeled by an amplitude equation of the form

$$\frac{\partial \Psi}{\partial t} = [\varepsilon - (\Delta + q_0)^2]\Psi - \Psi^3 . \tag{1}$$

This problem has received considerable attention also in laboratory experiments, employing both rectangular and cylindrical cells [10]. The numerical study of equation (1) provided the motivation for the work reported here.

2. Solution of the Biharmonic equation on a Disk

Consider the problem

$$\begin{aligned}
\Delta^2 u &= f & r &< R \\
u &= g & r &= R \\
u_r &= h & r &= R \ .
\end{aligned} \tag{2}$$

Glowinski and Pironneau [11] remarked that a discrete form of this problem derived from a finite difference grid based on polar coordinates, can be solved by using the "coupled equation approach", see [12]. This section describes a direct algorithm which is an order of magnitude faster. The method is based on the following characterization of biharmonic functions [13].

If $u(x,y)$ is biharmonic on a domain Ω and Ω is star shaped, then

$$u = r^2 v + w \tag{3}$$

where v and w are harmonic functions and $r^2 = x^2 + y^2$.

Let $u = u_1 + u_2$. First solve

$$\begin{aligned}
\Delta w_1 &= f & r &< R \\
w_1 &= 0 & r &= R
\end{aligned} \tag{4}$$

and then

$$\begin{aligned}
\Delta u_1 &= w_1 & r &< R \\
u_1 &= g & r &= R \ .
\end{aligned} \tag{5}$$

The problem for u_2 becomes

$$\begin{aligned}
\Delta^2 u_2 &= 0 & r &< R \\
u_2 &= 0 & r &= R \\
(u_2)_r &= h - (u_1)_r & r &= R \ .
\end{aligned} \tag{6}$$

Now write

$$u_2 = -(R^2 - r^2)v_1 + v_2 \tag{7}$$

and require that v_1 and v_2 be harmonic. Since v_2 vanishes at the boundary, it follows that it is identically zero. Now

$$\left(\frac{\partial u_2}{\partial r}\right)_{r=R} = 2Rv_1 \tag{8}$$

and therefore

$$\begin{aligned}
\Delta v_1 &= 0 & r &< R \\
v_1 &= \frac{1}{2R}(u_2)_r & r &= R \ .
\end{aligned} \tag{9}$$

In this way the numerical solution of (2) has been reduced to the solution of three Poisson equations on the same grid. The derivative $(u_1)_r$ which is needed in (6), must be computed with sufficient accuracy from the solution of (5). If a second order accurate method is used for solving Poisson's equation then the discrete approximation of $(u_1)_r$ should also be second order accurate. A second order accurate numerical solution to the original (smooth) problem can then be obtained.

A computer implementation of this idea using the subroutine PWSPLR [2] for the Poisson problems, has been written. The algorithm has an operation count of $O(NM \log N)$ when a discretization with N points in the θ-direction and M points in the r-direction is used. The code requires $2NM + O(N) + O(M)$ storage. A somewhat faster code requiring half the storage, could be implemented by taking advantage of the zero right hand side in (9). Numerical results are given in the last section of this paper.

3. Efficient solution of the generalized equation in an annulus

A linearization of problem (1) requires the solution of an equation of the form

$$
\begin{aligned}
\Delta^2 u + \alpha \Delta u + \beta u &= f && R_0 < r < R_1 \\
u &= g_1 && r = R_0 \\
u &= g_2 && r = R_1 \\
u_r &= h_1 && r = R_0 \\
u_r &= h_2 && r = R_1 \; .
\end{aligned}
\tag{10}
$$

In this equation α and β are scalar quantities. It is an open question whether the approach taken in the previous section, can be generalized to this case. Since the domains of interest also include the annulus ($R_0 < r < R_1$), we will now proceed to develop a direct algorithm for this case. We start out assuming $R_0 > 0$ and will include the case $R_0 = 0$ in the next section.

We introduce the following notation defining MN interior grid points:

$$
\begin{aligned}
\Delta r &= \frac{R_1 - R_0}{M+1} & r_i &= R_0 + i\Delta r & i &= 0,1,2,.. \; M+1 \\
\Delta \theta &= \frac{2\pi}{N} & \theta_j &= j\Delta\theta & j &= 0,1,2,.. \; N \; .
\end{aligned}
\tag{11}
$$

The five point stencil based on centered, second order accurate finite differences for the Laplace operator in polar coordinates has the coefficients given in (12). To be precise, the discrete equation for gridpoint (i,j) depends on the gridpoints indicated relative to (i,j) with the corresponding coefficients

$$
\begin{aligned}
(i,j): &\qquad -2\left(\frac{1}{\Delta r^2} + \frac{1}{(r_i\Delta\theta)^2}\right) \\[2ex]
(i,j+1): &\qquad \frac{1}{(r_i\Delta\theta)^2} \\[2ex]
(i,j-1): &\qquad \frac{1}{(r_i\Delta\theta)^2} \\[2ex]
(i+1,j): &\qquad \frac{r_{i+1/2}}{r_i\Delta r^2} \\[2ex]
(i-1,j): &\qquad \frac{r_{i-1/2}}{r_i\Delta r^2} \; .
\end{aligned}
\tag{12}
$$

The corresponding 13-point stencil for Δ^2 is given below.
First, the weight for gridpoint (i,j)

$$
(i,j): \qquad \frac{r_{i+1/2}^2}{r_i r_{i+1}\Delta r^4} + \frac{r_{i-1/2}^2}{r_i r_{i-1}\Delta r^4} + \frac{4}{\Delta r^4} + \frac{6}{(r_i\Delta\theta)^4} + \frac{8}{(r_i\Delta r\Delta\theta)^2} \; .
\tag{13}
$$

The coupling to points on the gridline having $r = R_0 + (i-1)\Delta r$ is

$$
\begin{aligned}
(i-1,j-1): &\qquad \frac{r_{i-1/2}}{r_i(\Delta r\Delta\theta)^2}\left(\frac{1}{r_i^2} + \frac{1}{r_{i-1}^2}\right) \\[2ex]
(i-1,j): &\qquad -2\frac{r_{i-1/2}}{r_i\Delta r^2}\left(\frac{2}{\Delta r^2} + \frac{1}{\Delta\theta^2}\left(\frac{1}{r_i^2} + \frac{1}{r_{i-1}^2}\right)\right) \\[2ex]
(i-1,j+1): &\qquad \frac{r_{i-1/2}}{r_i(\Delta r\Delta\theta)^2}\left(\frac{1}{r_i^2} + \frac{1}{r_{i-1}^2}\right) \; .
\end{aligned}
\tag{14}
$$

The coupling to neighbor points on the same gridline $(r = R_0 + i\Delta r)$ is

$$
\begin{aligned}
(i,j-2): &\quad \frac{1}{(r_i\Delta\theta)^4} \\[2mm]
(i,j-1): &\quad \frac{-4}{(r_i\Delta\theta)^2}\left(\frac{1}{\Delta r^2}+\frac{1}{(r_i\Delta\theta)^2}\right) \\[2mm]
(i,j+1): &\quad \frac{-4}{(r_i\Delta\theta)^2}\left(\frac{1}{\Delta r^2}+\frac{1}{(r_i\Delta\theta)^2}\right) \\[2mm]
(i,j+2): &\quad \frac{1}{(r_i\Delta\theta)^4} \;.
\end{aligned}
\tag{15}
$$

The coupling to points on gridline $r = R_0 + (i+1)\Delta r$ is

$$
\begin{aligned}
(i+1,j-1): &\quad \frac{r_{i+1/2}}{r_i(\Delta r\Delta\theta)^2}\left(\frac{1}{r_i^2}+\frac{1}{r_{i+1}^2}\right) \\[2mm]
(i+1,j): &\quad -2\frac{r_{i+1/2}}{r_i\Delta r^2}\left(\frac{2}{\Delta r^2}+\frac{1}{\Delta\theta^2}\left(\frac{1}{r_i^2}+\frac{1}{r_{i+1}^2}\right)\right) \\[2mm]
(i+1,j+1): &\quad \frac{r_{i+1}}{r_i(\Delta r\Delta\theta)^2}\left(\frac{1}{r_i^2}+\frac{1}{r_{i+1}^2}\right) \;.
\end{aligned}
\tag{16}
$$

Finally, the points having $r = R_0 + (i-2)\Delta r$ and $r = R_0 + (i+2)\Delta r$

$$
\begin{aligned}
(i-2,j): &\quad \frac{r_{i-1/2}\,r_{i-\frac{3}{2}}}{r_i r_{i-1}\Delta r^4} \\[3mm]
(i+2,j): &\quad \frac{r_{i+1/2}\,r_{i+\frac{3}{2}}}{r_i r_{i+1}\Delta r^4} \;.
\end{aligned}
\tag{17}
$$

In order to get a matrix description of our discrete equations we need some elementary matrices.

Let

$$
R_N \;=\; \begin{vmatrix}
2 & -1 & 0 & & . & -1 \\
-1 & . & . & & 0 & . \\
0 & . & . & & . & 0 \\
. & 0 & . & & . & -1 \\
-1 & & . & 0 & -1 & 2
\end{vmatrix}_{N\times N} \;.
\tag{18}
$$

The eigenvalues and eigenvectors of this matrix are well known,

$$
Q^T R Q \;=\; \Lambda
\tag{19}
$$

with $\lambda_1 = 0$ and $\lambda_N = 4$ if N is even. The corresponding orthonormal eigenvectors are

$$
q_1 \;=\; \sqrt{\frac{1}{N}}(1,1,\ldots 1)^T
\tag{20}
$$

$$
q_{2j} \;=\; \left|\sqrt{\frac{2}{N}}\cos\frac{2\pi j(i-1)}{N}\right|_{i=1}^{N}
\qquad j=1,2,\ldots\frac{N-1}{2}
\tag{21}
$$

$$
q_{2j+1} \;=\; \left|\sqrt{\frac{2}{N}}\sin\frac{2\pi j(i-1)}{N}\right|_{i=1}^{N}
\qquad j=1,2,\ldots\frac{N-1}{2} \;.
\tag{22}
$$

If N is even then

$$
q_N \;=\; \sqrt{\frac{1}{N}}(1,-1,1,\ldots 1,-1)^T \;.
\tag{23}
$$

Notice that the product

$$y = Qx \qquad or \qquad x = Q^T y \tag{24}$$

can be computed using the fast Fourier transform algorithm. The first product (Qx) is an inverse real FFT while the second ($Q^T y$) corresponds to a forward real transform. If the vector has N components then the required work is of the order $N\,logN$. Also notice that if y is a vector with all components equal to 1 , then x is the (scaled) unit vector with $\sqrt{N}$ in the first position.

The discrete Laplacian $L = \Delta_{\Delta r \Delta \theta}$ can now be written

$$L = \frac{1}{\Delta r^2}(I \otimes \widetilde{A}) - \frac{1}{\Delta \theta^2}(R \otimes D^2) \tag{25}$$

with

$$\widetilde{A} = tridiagonal(\widetilde{a}_i, -2, \widetilde{b}_i)$$

$$D = diagonal(\frac{1}{r_i})$$

$$\widetilde{a}_i = \frac{r_{i-1/2}}{r_i} \quad , \quad \widetilde{b}_i = \frac{r_{i+1/2}}{r_i} \ . \tag{26}$$

Let

$$A = D^{-1/2}\widetilde{A}D^{1/2} \ . \tag{27}$$

Then A is tridiagonal and symmetric. The diagonal term is -2, the off-diagonal term a_i is

$$a_i = \frac{1}{2} \frac{r_i + r_{i+1}}{\sqrt{r_i r_{i+1}}} \quad ,i = 1,2,...m-1. \tag{28}$$

Using this, we obtain the formula

$$L^2 = \frac{1}{\Delta r^4}(I \otimes \widetilde{A}^2) - \frac{1}{(\Delta r \Delta \theta)^2}(R \otimes \widetilde{A}D^2 + R \otimes D^2\widetilde{A}) + \frac{1}{\Delta \theta^4}(R^2 \otimes D^4) \ . \tag{29}$$

In order to describe the complete discrete system of equations for the biharmonic problem, we need modifications to the stencil (13-17) for points that are only a distance Δr from the inner or outer boundary. This can formally be described by the introduction of extra gridpoints in the exterior of the domain, a distance Δr from the boundary. By using the normal derivative boundary condition, extrapolated values at these point can be used in the original stencil, making it well defined at all interior gridpoints. The net result of this is a modification of the matrix coefficients corresponding to the gridpoints a distance Δr from the boundary. For the inner boundary $r = R_0$ we get

$$u_{-1j} = u_{1j} - 2\Delta r u_r + O(\Delta r^3) \qquad j = 1,2,...N \ . \tag{30}$$

Notice that this results in a much less accurate approximation to the differential operator at these points, but it can be shown [14] that the discrete solution is second order accurate. The term $2\Delta r u_r$ with a weight according to the stencil given in (17) will contribute to the right hand side of the discrete set of equations. From (30) we derive the modification of the matrix coefficient given by (29). A simple calculation shows that the correct change is obtained by adding $\frac{2}{\Delta r^4}\widetilde{a}_1$. Similarly, we get the correction term $\frac{2}{\Delta r^4}\widetilde{b}_M$ to the matrix coefficients corresponding to gridpoints u_{Mj} ,$j = 1,2,...N$ at the boundary $r = R_1$.

Let the $M \times 2$ matrix U be zero except for the element $\sqrt{\widetilde{a}_1}$ in the (1,1) position and the element $\sqrt{\widetilde{b}_M}$ in the (M,2) position. We can now formulate a discrete approximation of problem (10)

$$Bx = b \ . \tag{31}$$

In this problem having MN unknowns, b contains the discrete data from (10), incorporating the boundary values in the way specified by the 13-point stencil and outlined above. This equation can equivalently be written

$$S y = c \tag{32}$$

where

$$S = (Q^T \otimes I)(I \otimes D^{-1/2}) B (I \otimes D^{1/2})(Q \otimes I) \tag{33}$$

$$x = (I \otimes D^{1/2})(Q \otimes I) y \tag{34}$$

$$c = \Delta\theta^4 (Q^T \otimes I)(I \otimes D^{-1/2}) b . \tag{35}$$

An easy calculation will show that S is block diagonal, each block symmetric and penta-diagonal. The elements of S are all explicitly known from the formulas given earlier.

$$S = (\frac{\Delta\theta}{\Delta r})^4 (I \otimes A^2) - (\frac{\Delta\theta}{\Delta r})^2 (\Lambda \otimes (AD^2 + D^2 A) + \Lambda^2 \otimes D^4 + \tag{36}$$

$$+ \alpha\Delta\theta^2((\frac{\Delta\theta}{\Delta r})^2 (I \otimes A) - \Lambda \otimes D^2) + \beta\Delta\theta^4 (I \otimes I) + 2(\frac{\Delta\theta}{\Delta r})^4 (I \otimes (D^{-1/2} U)(D^{1/2} U)^T) .$$

The description of a solution algorithm for this problem that only requires $O(MN \log N)$ arithmetic operations and $MN + O(M) + O(N)$ storage locations, is therefore complete.

It should be noticed that this solution is a straight forward implementation of classical 'fast Poisson solver' techniques. The coordinate system makes the expressions more involved, but the idea is simple. If the problem is generalized allowing the domain to be a sector of the annulus, (with four boundary segments) then the nontrivial method described in [8] for the same problem in rectangular coordinates, could be implemented. This would again lead to a computationally optimal algorithm.

4. The case $R_0 = 0$

The stencils for the Laplace and biharmonic operators (12) and (13-17), clearly breaks down in the case $R_0 = 0$. Consider first the approximation of Δu at the point $r = 0$. An approximation of the form

$$(\Delta u)_{r=0} \approx a\, u_{00} + \frac{b}{N} \sum_{j=1}^{N} u_{1j} \tag{37}$$

results in $a = -b$ with $b = (\frac{2}{\Delta r})^2$ if we require the formula to be exact for $u = 1$ and $u = r^2$. An alternative derivation can be found in [2].

Proceeding in this way, we arrive at the formula

$$(\Delta^2 u)_{r=0} \approx (\frac{2}{\Delta r})^4 (u_{00} - \frac{4}{3N} \sum_{j=1}^{N} u_{1j} + \frac{1}{3N} \sum_{j=1}^{N} u_{2j}) . \tag{38}$$

This approximation is exact for $u = 1$, $u = r^2$ and $u = r^4$. We will use this expression as the equation for the center point. Since the finite difference stencil extends over two levels of grid points in all directions, modified approximations are also required for the grid points on the innermost circle at a distance Δr from the center point. A straight forward derivation shows that the points $(i-1, j-1)$, $(i-1, j+1)$ and $(i-2, j)$ disappear (as they must), the center point enter with a weight $-\frac{3}{\Delta r^4}$, and all the points on the innermost circle u_{1k}, $k = 1, 2, \ldots N$ enter with the weight $\frac{2}{N} \frac{1}{\Delta r^4}$. Finally, if we consider the equation for gridpoint $(1, j)$, its coefficient is modified removing the singular term

$$(1, j): \quad \frac{6}{\Delta r^4 \Delta\theta^4} + \frac{8}{\Delta r^4 \Delta\theta^2} + \frac{41}{8\Delta r^4} . \tag{39}$$

It should be noticed that these approximations are less accurate than the general 13-point formula, when viewed as an approximation of the differential operator. However, the solution of the discrete equations will give a second order accurate approximation to the unknown function u. The situation is in this respect similar to the procedure that must be used near the boundary.

The coefficient matrix for the new system becomes

<table>
<tr><td>a</td><td>y^T</td></tr>
<tr><td>x</td><td>C</td></tr>
</table>

where

$$C = B + \frac{2}{\Delta r^4}\frac{1}{N}(ee^T\otimes e_1 e_1^T) \ . \tag{40}$$

Here e_1 is the first unit vector and e is a vector having all components equal to one. The transformation

$$Q^T e = \sqrt{N} e_1 \tag{41}$$

preserves the structure of the transformed matrix as in the previous case. This is a key observation making it easy to solve linear systems with the matrix C. The vectors x and y augmenting the matrix C are both of the form

$$x = (e\otimes\tilde{x}) \qquad y = (e\otimes\tilde{y}) \tag{42}$$

and $\tilde{x}$ and $\tilde{y}$ have nonzeros in the two first positions only. We get the following modified linear system

$$a\,u_{00} + y^T u = f_{00} \tag{43}$$
$$u_{00}x + Cu = f \ .$$

This system is solved by first computing u_{00} from

$$(a - y^T C^{-1} x)\,u_{00} = f_{00} - y^T C^{-1} f \tag{44}$$

and then

$$u = C^{-1} f - u_{00} C^{-1} x \ . \tag{45}$$

We need to solve $Cv = f$ and $Cz = x$, but from equation (41) it follows that only the first block, in the block diagonal system that C transforms to, has a nonzero right hand side. Going back, we get z of the form

$$z = (e\otimes\tilde{z}) \tag{46}$$

since y also is of this form, the essential work is to solve $Cv = f$. The additional work is an order of magnitude smaller. The complexity of the algorithm described in section 3, is therefore unaltered when the important, special case considered in this section, is included.

5. Numerical Results

The algorithm described in the previous sections have been implemented as Fortran subroutines. (They are available together with codes for the rectangular domain from the author.) The first table contains the results from solving equation (10) with $\alpha = 2$ and $\beta = 7$ for a problem with exact solution $u = e^{r\sin\theta}$. The time refers to cpu-time on a VAX-11/780 with floating point accelerator. The number of gridpoints in each coordinate direction is N. The code does not require N to be a power of 2. It is clear from the successive ratios of both the L_2 errors and the maximum errors, that the solution is second order accurate.

N	Time	L_2	Ratio	Max	Ratio
4	0.04	3.0 (-2)	-	2.5 (-2)	-
8	0.08	7.8 (-3)	2.51	9.8 (-3)	3.90
16	0.26	2.0 (-3)	3.88	2.5 (-3)	3.88
32	1.05	5.1 (-4)	3.96	6.4 (-4)	3.96
64	4.08	1.3 (-4)	3.98	1.6 (-4)	3.98
128	16.60	3.2 (-5)	3.99	4.0 (-5)	3.99
256	67.12	8.0 (-6)	4.00	1.0 (-5)	4.00
512	294.52	2.0 (-6)	3.93	2.6 (-6)	3.97

Table 1.

The next table shows a problem with the same exact solution, but $\alpha = \beta = 0$ thus making it possible to try the algorithm developed in section 2. Notice that the accuracy is very similar, but with a significantly longer computing time. This code could be optimized to some degree, but the main work is the three Poisson solutions. The results indicate that the algorithm developed in section 3 and 4 is about as fast as an efficient algorithm for Poissons equation on the same domain.

N	Time	L_2	Ratio	Max	Ratio
4	0.04	3.2 (-2)	-	2.2 (-2)	-
8	0.14	7.6 (-3)	4.2	8.2 (-3)	2.6
16	0.62	2.0 (-3)	3.9	2.1 (-3)	3.9
32	2.90	4.9 (-4)	3.9	5.3 (-4)	4.0
64	14.15	1.2 (-4)	4.0	1.3 (-4)	4.0
128	62.26	3.1 (-5)	4.0	3.3 (-5)	4.0

Table 2.

Acknowledgements

I would like to thank H.S. Greenside, W.M. Coughran,Jr. and N. Schryer for the motivation to complete this work.

References

1. A.C. Newell and J.A. Whitehead, Finite Bandwidth,
   Finite Amplitude Convection, Journal of Fluid Mechanics
   38, 279-303 1969.

2. P.N. Swarztrauber and R. Sweet, Efficient Fortran subprograms
   for the solution of elliptic partial differential equations.
   NCAR-TN/IA-109. National Center for Atmospheric Research,
   Boulder, Colorado, 1975.

3. A. Brandt, Guide to multigrid development, Multigrid Methods
   (W. Hackbusch, U. Trottenberg eds.), Lecture Notes in
   Mathematics. Springer-Verlag, Berlin 1982.

4. K. Stüben and U. Trottenberg, Multigrid methods: Fundamental
   algorithms, model problem analysis and applications,
   Multigrid Methods (W. Hackbusch, U. Trottenberg eds.),
   Lecture Notes in Mathematics. Springer-Verlag, Berlin 1982.

5. W. Proskurowski and O. Widlund, On the numerical solution
   of Helmholtz's equation by the capacitance method.
   Math. Comp. Vol. 30, 433-468 1976.

6. W. Proskurowski and O. Widlund, A finite element capacitance
   matrix method for the Neuman problem for Laplace's equation.
   SIAM J. Sci. Stat. Comput. Vol. 1, 410-425 1980.

7. P.E. Bjørstad and O.B. Widlund, Solving Elliptic
   Problems on Regions Partitioned into Substructures.
   Proceedings of the Elliptic Problem Solvers Meeting, held in
   Monterey, California January 10-12 1983. Academic Press 1984.

8. P.E. Bjørstad, Fast Numerical Solution of the
   Biharmonic Dirichlet Problem on Rectangles.
   SIAM Journal on Numerical Analysis 20, 59-71 1983.

9. P.E. Bjørstad, W.M. Coughran Jr., H.S. Greenside, D.J. Rose,
   and N.L. Schryer, Numerical solution of a model equation
   near the onset of the Rayleigh-Bénard instability.
   Proceedings of the Elliptic Problem Solvers Meeting, held in
   Monterey, California January 10-12 1983. Academic Press 1984.

10. H.S. Greenside, W.M. Coughran, Jr. and N.L. Schryer,
    Nonlinear pattern Formation Near the Onset of
    Rayleigh-Benard Convection.
    Phys. Rev. Lett. Vol. 49, 726-729, 1982.

11. R. Glowinski and O. Pironneau, Numerical methods
    for the first biharmonic equation and for the
    two-dimensional Stokes problem.
    SIAM Review. Vol. 21, 167-212, 1979.

12. J.W. McLaurin, A general coupled equation approach
for solving the biharmonic boundary value problem.
SIAM J. Numer. Anal. Vol 11, 14-33, 1974.

13. A.N. Tychonoff and A.A. Samarski, Differentialgleichungen
der Matematischen Physik, pp 389.
VEB Deutscher Verlag der Wissenschaften, Berlin 1959.

14. P.E. Bjørstad, Numerical Solution of the Biharmonic Equation.
Ph.D. Dissertation, Stanford University 1980.

ON THE STABILIZATION OF FINITE ELEMENT
APPROXIMATIONS OF THE STOKES EQUATIONS

F. Brezzi[1] and J. Pitkäranta[2]

ABSTRACT

Consider finite element approximation of the Stokes equations.
We present a systematic way of stabilizing it by adding bubble
functions to the discrete velocity field. Another way of
stabilization is also presented where the finite element spaces
are kept unchanged but the discrete incompressibility condition
is modified instead.

1. INTRODUCTION

Assume for the sake of simplicity that Ω is a given poly-
hedron in $\mathbb{R}^n$ $(n \geq 2)$ and set $V = (H_0^1(\Omega))^n$, $P = L^2(\Omega)$;
consider now for a given $\underline{f}$, say, in $(L^2(\Omega))^n$ and $g \in P/\mathbb{R}$
the (generalized) Stokes problem

$$
\begin{cases}
\text{Find } \underline{u} \in V, p \in P \text{ such that:} \\[2ex]
\text{(i)} \quad a(\underline{u},\underline{v}) - \int_\Omega p \operatorname{div} \underline{v} \, dx = \int_\Omega \underline{f} \cdot \underline{v} \, dx \quad \forall \underline{v} \in V, \\[2ex]
\text{(ii)} \quad \int_\Omega q \operatorname{div} \underline{u} \, dx = \int_\Omega gq \, dx \quad \forall q \in P
\end{cases}
\tag{1.1}
$$

where, as usual

$$
a(\underline{u},\underline{v}) := \int_\Omega \sum_{r=1}^n \sum_{i=1}^n \frac{\partial u_r}{\partial x_i} \frac{\partial v_r}{\partial x_i} \, dx.
\tag{1.2}
$$

[1] Dipartimento di Meccanica Strutturale - Univ. of Pavia and
Institutio di Analisi Numerica del C.N.R. - 27100 Pavia (Italy)
[2] Institute of Mathematics, Helsinki University of Technology, SF-02150 Espoo 15, Finland

It is well known that problem (1.1) has a unique solution. Consider now a finite element discretization of (1.1) consisting (as usual) of two families $\{V_h\},\{P_h\}$ of finite dimensional subspaces of V and P respectively, and the corresponding discretized problems

$$
\left\{
\begin{array}{l}
\text{Find } \underline{u}_h \in V_h, \ p_h \in P_h \quad \text{such that:} \\[2ex]
a(\underline{u}_h,\underline{v}_h) - \int_\Omega p_h \operatorname{div} \underline{v}_h \, dx = \int_\Omega \underline{f} \cdot \underline{v}_h \, dx \qquad \forall \underline{v}_h \in V_h, \qquad (1.3) \\[2ex]
\int_\Omega q_h \operatorname{div} \underline{u}_h \, dx = \int_\Omega g q_h \, dx \qquad \forall q_h \in P_h .
\end{array}
\right.
$$

It is also well known that one is not allowed to take independent choices for V_h and P_h in order to have stability and convergence results (c.f. [2], [3]). In the present paper we consider the possibility of modifying an unstable scheme with a minimal cost so that stability can be recovered.

In the first case we show that by adding suitable bubble functions to the spaces V_h, one can stabilize "any" given pair of families $\{V_h\},\{P_h\}$. More precisely, taking for instance triangular elements, this can be done for any choice of $\{P_h\}$ provided that the spaces $\{V_h\}$ contain at least all piecewise quadratic continuous functions (vanishing on $\partial\Omega$); this can also be done under the weaker assumption that the $\{V_h\}$'s contain at least all piecewise linear continuous functions (vanishing on $\partial\Omega$) provided that $P_h \subseteq H^1(\Omega)$ for all h. The choice of bubble functions (that is: functions whose support is contained in a single element) as stabilizing correction relies on the fact that one can eliminate (i.e., condense) such degrees of freedom without affecting the structure of the stiffness matrix. Thus if direct solution techniques are used, the bubble functions have only a minor contribution to the total computational cost.

We consider also another way of achieving stability where we do not modify the discrete velocity or pressure fields but instead modify the discrete equations by adding a stabilizing

term into the discrete equations. Unlike in the bubble function approach, this type of stabilization affects also the consistency of the scheme. However, if the extra consistency
error is not larger than that of the original scheme, this
approach may be useful when iterative (such as multigrid)
techniques are used in the solution of the linear system.

Stabilization by adding bubble functions is discussed in Section 2 below. The proof in the case $P_h \subset H^1(\Omega)$ is essentially
contained in [1], but we report it below for the sake of
completeness. In Section 3 we give a simple example of
stabilization by modifying the discrete equations. Another
example is given in [7], where the same type of stabilization
is applied to the "bilinear velocities - constant pressure"
approximation of the Stokes problem.

2. STABILIZATION WITH BUBBLE FUNCTIONS

For the sake of simplicity we consider only the case of
"triangular" elements: however it will be clear from the proofs
that the results hold for "quadrilateral" elements as well,
and also for more general discretizations.

Let $\{C_h\}$ be a family of decompositions of Ω into n-simplexes
and let $\{V_h\}, \{P_h\}$ be the corresponding given families of
f.e.m. Let moreover, for any n-simplex $K, b_K(x)$ be the
corresponding bubble function of degree $n+1$:

$$b_K(x) = c\lambda_1(x)\lambda_2(x) \ldots \lambda_{n+1}(x) \tag{2.1}$$

where the λ_i's are the barycentric coordinates ($\equiv$ equations
of the faces) and c is a normalizing factor so that, say

$$\sup_K \, b_K(x) = 1; \tag{2.2}$$

note that $b_K(x) > 0$ in the interior of K. We are looking
for corrections of V_h of type

$$\begin{cases} \tilde{V}_h = V_h \oplus B_h \; ; \qquad B_h = \underset{K \in C_h}{\oplus} B_h^K \; ; \\[2ex] B_h^K = \begin{cases} b_K \nabla P_h|_K & \text{in } K \\[1ex] 0 & \text{elsewhere;} \end{cases} \end{cases} \qquad (2.3)$$

clearly $B_h \subseteq (H_0^1(\ ))^n$; we want to show that the corresponding modified f.e.m:

$$\begin{cases} \text{Find } \tilde{\underline{u}}_h \in \tilde{V}_h, \; p_h \in P_h \quad \text{such that:} \\[2ex] a(\tilde{\underline{u}}_h, \tilde{\underline{v}}_h) - \int_\Omega p_h \operatorname{div} \tilde{\underline{v}}_h \, dx = \int_\Omega \underline{f} \cdot \tilde{\underline{v}}_h \, dx \quad \forall \tilde{\underline{v}}_h \in \tilde{V}_h \\[2ex] \int_\Omega q_h \operatorname{div} \tilde{\underline{u}}_h \, dx = \int_\Omega g q_h \, dx \quad \forall q_h \in P_h \end{cases} \qquad (2.4)$$

is stable and gives "optimal" error bounds:

$$\begin{aligned} \|\underline{u} - \tilde{\underline{u}}_h\|_1 + \|p - p_h\|_{0/\mathbb{R}} \leq c\{ &\inf_{\tilde{v}_h \in \tilde{V}_h} \|u - \tilde{v}_h\|_1 + \\ &+ \inf_{q_h \in P_h} \|p - q_h\|_{0/\mathbb{R}} \}, \end{aligned} \qquad (2.5)$$

with c independent of h. As is well known (c.f. [2], [3]), the stability condition required for (2.5) to hold is: there is a constant C independent of h, f and g such that $(\tilde{u}_h, p_h)$ satisfies

$$\|\tilde{u}_h\|_1 + \|p_h\|_{0/\mathbb{R}} \leq C(\|f\|_{-1} + |g|_{0/\mathbb{R}}) \qquad (2.6)$$

where $\|\cdot\|_{-1}$ denotes the dual norm of $(H_0^1(\Omega))^2$. It is also known that a sufficient condition in order to have (2.6) is (c.f. [3], [6])

$$\begin{cases} \exists \tilde{\pi}_h \in L(V, \tilde{V}_h) \quad \text{such that} \\[2ex] \text{i) } \|\tilde{\pi}_h\| \leq c \quad \text{(indep. of } h\text{)} \quad \text{in } L(V,V) \\[2ex] \text{ii) } \int_\Omega q_h \operatorname{div}(\underline{v} - \tilde{\pi}_h \underline{v}) \, dx = 0 \quad \forall \underline{v} \in V, \; \forall q_h \in P_h \; . \end{cases} \qquad (2.7)$$

The following two theorems give sufficient conditions on V_h so that $\tilde{V}_h$, as defined in (2.3), satisfy (2.7) (and hence (2.5)).

THEOREM 1. <u>Assume that</u>:

$$\left\{ \begin{array}{l} \exists \pi_h \in L(V,V_h) \quad \text{s.t.} \quad \sum_K h_K^{2r-2} \|\underline{v}-\pi_h\underline{v}\|_r^2 \le c\|\underline{v}\|_1^2 \\[2ex] \underline{\text{for}} \quad r = 0,1, \;\underline{\text{with}} \quad c \quad \underline{\text{independent of}} \quad h \\[1ex] \underline{\text{where}} \quad h_K = \underline{\text{diameter of}} \quad K, \end{array} \right. \tag{2.8}$$

$$P_h \subseteq H^1(\Omega); \tag{2.9}$$

<u>then the pair</u> $\tilde{V}_h, P_h$ <u>obtained from</u> (2.3) <u>satisfies</u> (2.5).

PROOF. We look for $\tilde{\pi}_h$ of the form

$$\tilde{\pi}_h\underline{v} = \pi_h\underline{v} + \underline{\beta}_h(\underline{v}), \quad \underline{\beta}_h \in B_h. \tag{2.10}$$

Then (2.7 ii) is satisfied if, for all K in $\mathcal{C}_h$,

$$\int_K (\underline{v}-\pi_h\underline{v}-\underline{\beta}_h) \cdot \nabla q_h = 0 \qquad \forall q_h \in P_h. \tag{2.11}$$

This uniquely determines $\underline{\beta}_h \in B_h^K$. Moreover one has

$$\|\beta_h\|_{1,K} \le ch_K^{-1}\|\underline{v}-\pi_h\underline{v}\|_{0,K} \tag{2.12}$$

so that (2.8), (2.10) and (2.12) give (2.7 i).

COROLLARY. <u>If</u> $P_h \subseteq H^1(\Omega)$ <u>and</u> V_h <u>contains at least all piecewise linear continuous functions vanishing on</u> $\partial\Omega$, <u>then</u> $\tilde{V}_h, P_h$ <u>satisfies</u> (2.5).

PROOF. It is clear (see e.g. [1], [4]) that the present assumption on V_h implies (2.8).

THEOREM 2. <u>Assume that</u> $\exists \pi_h \in L(V,V_h)$ <u>such that</u>:

$$\int_K \text{div}(\underline{v}-\pi_h\underline{v})\,dx = 0 \qquad \forall K, \ \forall \underline{v} \in V \tag{2.13}$$

$$\|\pi_h\| \leq c \quad \text{independent of} \ \ h \tag{2.14}$$

then the pair $\tilde{V}_h, P_h$ obtained from (2.3), satisfies (2.5).

PROOF. We look again for $\tilde{\pi}_h$ of the form (2.10). Now for any given $q_h \in P_h$ we consider the decomposition $q_h = q_h^{(1)} + q_h^{(2)}$ with $q_h^{(1)} =$ piecewise constant and $\int_K q_h^{(2)}\,dx = 0$ for all K. Then

$$\int_K \text{div}(\underline{v}-\tilde{\pi}_h\underline{v})q_h\,dx = \int_K \underline{\beta}_h \cdot \nabla q_h\,dx + \int_K \text{div}(\underline{v}-\pi_h\underline{v})q_h^{(2)}\,dx \tag{2.15}$$

so that (2.7 ii) holds if

$$\int_K \underline{\beta}_h \cdot \nabla q_h\,dx = - \int_K q_h^{(2)}\,\text{div}(\underline{v}-\pi_h\underline{v})\,dx \tag{2.16}$$

which again determines uniquely $\underline{\beta}_h \in B_h$. Since $q_h^{(2)}$ has locally zero mean value, a simple scaling argument shows that

$$\|\underline{\beta}_h\|_1 \leq c\|\text{div}(\underline{v}-\pi_h\underline{v})\|_0 \leq c|\underline{v}-\pi_h\underline{v}|_1 \tag{2.17}$$

with c independent of h. Now (2.14), (2.10) and (2.17) give (2.7 i).

COROLLARY. If V_h contains at least all piecewise quadratic continuous functions vanishing on $\partial\Omega$, then $\tilde{V}_h, P_h$ satisfies (2.5).

PROOF. The result follows immediately from the fact that the "quadratic velocities - constant pressure" approximation of the Stokes problem satisfies (2.6) (c.f. [5]): hence for any given $\underline{v} \in V$ we may choose $\pi_h\underline{v}$ as the discretized solution of $-\Delta\underline{u} + \Delta p = -\Delta\underline{v}$ and $\text{div}\,\underline{u} = \text{div}\,\underline{v}$ (that is $\underline{u} = \underline{v}$, $p = 0$), by means of the quadratic-constant scheme. In our assumptions we have $\pi_h\underline{v} \in V_h$ and hence (2.13), (2.14) are satisfied.

REMARK. With a slightly different proof we see that we could also deal with nonconforming V_h; in that case we could accept, for any choice of P_h, to start with spaces V_h that contain at least P_1-nonconforming spaces like in [5].

3. STABILIZATION BY MODIFYING THE DISCRETE EQUATIONS

To present the idea, let us consider a simple example where P_h is the space of piecewise linear continuous functions associated to a triangulation C_h of a two-dimensional polygonal domain, and $V_h = P_h^2 \cap H_0^1(\Omega)$. The pair (V_h, P_h) is not stable in the sense of (2.7), but we can modify the discrete equations so that the scheme becomes stable in the sense of (2.6). To this end, let us define the approximate solution to (1.1) as the pair $(\underline{u}_h, P_h)$ which satisfies (1.3 i) and

$$\sum_{K \in C_h} h_K^2 \int_K \nabla P_h \cdot \nabla q_h \, dx + \int_\Omega q_h \, \mathrm{div}\, \underline{u}_h \, dx$$

$$= \int_\Omega g q_h \, dx \qquad \forall q_h \in Q_h \quad . \tag{3.1}$$

THEOREM 3. <u>If</u> $(\underline{u}_h, P_h)$ <u>satisfies</u> (1.3i), (3.1) <u>then</u> (2.6) <u>holds</u> (with $\tilde{u}_h := u_h$) <u>and one has the error estimate</u>

$$\|\underline{u} - \underline{u}_h\|_1 + \|p - p_h\|_{0/R} \le Ch(|\underline{u}|_2 + |p|_1). \tag{3.2}$$

REMARK. The error estimate does not follow directly from (2.6) because we have modified (1.3 ii) thus affecting the consistency of the scheme. In other words, we need a separate estimate for the extra consistency error caused by the added stabilizing term in (3.1).

PROOF. Let the triangles of C_h be grouped together to form disjoint polygonal "macroelements" each containing at most a fixed number, say N, triangles of C_h. Denote the coarse partioning of Ω so obtained by $\tilde{C}_h$ and let $\tilde{P}_h$ be the space of functions which are piecewise constant on the macroelements. Now it is well known (c.f. [5], [8]) that by a suitable choice of $\tilde{C}_h$ the pair $(V_h, \tilde{P}_h)$ is stable in the sense of (2.7), i.e., for any $\tilde{p}_h \in \tilde{P}_h$ there exists $\underline{w}_h \in V_h$ such that

$$\begin{cases} (\tilde{p}_h, \text{div } \underline{w}_h) \geq \|\tilde{p}_h\|^2_{0/\mathbb{R}} \\[2ex] \|\underline{w}_h\| \leq C\|\tilde{p}_h\|_{0/\mathbb{R}} \; . \end{cases} \tag{3.3}$$

For example, if C_h is obtained from a coarser triangulation C_h^0 by subdividing each triangle into four equal subtriangles, then it suffices to take $\tilde{C}_h = C_h^0$.

Now with u_h, p_h satisfying (1.3 i) and (3.1) define $\tilde{p}_h \in \tilde{P}_h$ as the local average of p_h on each macroelement. Then setting $\underline{v}_h = \underline{u}_h - \delta \underline{w}_h$ and $q_h = p_h$ in (1.3 i) and (3.1), with δ a constant and $\underline{w}_h$ satisfying (3.3), we obtain

$$a(\underline{u}_h, \underline{u}_h) + \delta \|\tilde{p}_h\|^2_{0/\mathbb{R}} + \sum_{T \in C_h} h_K^2 \int_K |\nabla p_h|^2 dx$$

$$- \delta a(\underline{u}_h, \underline{w}_h) + \delta(p_h - \tilde{p}_h, \text{div } \underline{w}_h) = (\underline{f}, \underline{v}_h) + (g, p_h) .$$

Now for any macroelement $M \in \tilde{C}_h$ we have the inequalities

$$C^{-1} \int_M |p_h - \tilde{p}_h|^2 dx \leq \sum_{\substack{K \in C_h \\ K \subseteq M}} h_K^2 \int_K |\nabla p_h|^2 dx \leq C \int_M |p_h - \tilde{p}_h|^2 dx \tag{3.4}$$

where C depends only on N and on the minimal angle of the triangles contained in M, so we may assume C to be an absolute constant. Choosing then a sufficiently small positive value for the parameter δ we have combining (3.3) and (3.4) that

$$\|\underline{u}_h\|^2_1 + \|p_h\|^2_{0/\mathbb{R}} \leq C\{(\underline{f}, \underline{v}_h) + (g, p_h)\} . \tag{3.5}$$

Applying on the right side the obvious estimate $\|\underline{v}_h\|_1 \leq C(\|\underline{u}_h\|_1 + \|p_h\|_{0/\mathbb{R}})$, the asserted stability estimate (2.6) follows.

Using the stability estimate we obtain by usual manipulations an error bound of the form

$$\|\underline{u}-\underline{u}_h\|_1 + \|p-p_h\|_{0/\mathbb{R}}$$

$$\leq C \inf_{(\underline{v}_h,q_h)\in V_h \times P_h} \{\|\underline{u}-\underline{v}_h\|_1 + \|p-q_h\|_{0/\mathbb{R}}$$

$$+ \sup_{\substack{r_h \in P_h \\ \|r_h\|_{0/\mathbb{R}}=1}} \sum_{K\in C_h} h_K^2 \int_K \nabla q_h \cdot \nabla r_h \, dx\}$$

where the last term on the right side arises from the added stabilizing term in (3.1). Since $\sum_K h_k^2 \int_K |\nabla r_h|^2 dx \leq C\|r_h\|_{0/\mathbb{R}}^2$

for $r_h \in P_h$, (3.2) follows by choosing $(\underline{v}_h,q_h)$ to be an appropriate interpolant of $(\underline{u},p)$.

REFERENCES

[1] D.N. Arnold, F. Brezzi, M. Fortin, A stable finite element for the Stokes equations, Preprint (1983), University of Pavia.

[2] I. Babuška, Error bounds for the finite element method, Numer. Math. 16 (1971), pp. 322-333.

[3] F. Brezzi, On the existence uniqueness and approximation of saddle point problems arising from Lagrange multipliers, R.A.I.R.O Anal. Numér. 2 (1974), pp. 129-151.

[4] P. Clement, Approximation by finite element functions using local regularization, R.A.I.R.O Anal. Numér. 9 R-2 (1975), pp. 77-84.

[5] M. Crouzeix, P.-A. Raviart, Conforming and non-conforming finite element methods for solving the stationary Stokes equations, R.A.I.R.O Anal. Numér. 7 R-3 (1977), pp. 33-76.

[6] M. Fortin, An analysis of the convergence of mixed finite element methods, R.A.I.R.O Anal. Numér. 11 R-3 (1977), pp. 341-354.

[7] J. Pitkäranta, A multigrid version of a simple finite element method for the Stokes problem, to appear.

[8] R. Stenberg, Analysis of mixed finite element methods for the Stokes problem: a unified approach, to appear in Math. Comp.

A MULTIGRID METHOD FOR CAUCHY-RIEMANN AND STEADY EULER EQUATIONS BASED ON FLUX-DIFFERENCE SPLITTING

E. DICK

Department of Machinery, State University of Ghent

Sint Pietersnieuwstraat 41, 9000 Gent, Belgium

SUMMARY

Flux-vector splitting and flux-difference splitting techniques are applied to Cauchy-Riemann equations. It is shown that for both techniques efficient multigrid methods can be constructed based on relaxation algorithms. The flux-difference splitting technique is applied to steady one-dimensional Euler equations and the resulting set of discrete equations is solved by a relaxation algorithm. The solution for transonic flow is free of transition points in the shock region. By analogy with the Cauchy-Riemann equations, it is concluded that this technique is extendable to two dimensions and that it can be used in the multigrid method.

INTRODUCTION

In recent years, the use of upwind finite difference schemes for solving Euler equations has gained considerable popularity. This is mainly due to the introduction of the flux-vector splitting approach by Steger and Warming [1] which has made it clear how concepts from the theory of characteristics, already known for a long time for quasi-linear first order hyperbolic systems [2], and further developed by many workers (see references in [1]), can be used on systems of conservation laws.

Up to now, the flux-vector splitting approach for calculating steady inviscid flow, described by Euler equations, mainly has been used in time-marching techniques of both explicit and implicit form [1,3,4,5]. Only very recently, it has been shown by Jespersen [6] that the flux-vector splitting approach also can be used directly on steady Euler equations to generate discrete equations that are amenable to a solution by relaxation methods, possessing smoothing properties so that they can be used in the multigrid technique, resulting in a very high computational efficiency.

The flux-vector splitting technique, in its finite difference form used by Jespersen, has some difficulties with the treatment of shocks. Since the technique relies on the use of forward and backward difference operators, on a non-uniform grid a coordinate transformation is necessary, introducing metrics inside the differential operators. Since the forward and backward difference operators are used in the same point, the conservation laws can not be satisfied exactly. This generates inaccuracies near sonic lines and typical under- and overshoots in shock regions. In order to alleviate problems of this kind, also encountered in time marching techniques [3,4], the flux vector splitting approach was extended to a control volume formulation by Deese [5], leaving out problems with metrics. However, due to difficulties in defining the split fluxes between cells having different sign of one of the eigenvalues of the system of equations, still some oscillations in shock regions are found.

The question of shock treatment was studied thoroughly by Osher and Chakravarthy [7] and by Van Leer [8]. The conclusion was that a flux-vector splitting technique can at best represent a shock with the introduction of two transition points.

A technique, dual to flux-vector splitting, called flux-difference splitting was introduced by Roe [9], based on the earlier work of Godunov [10]. In this technique, not the flux-vectors but the flux-differences between opposite surfaces of a control volume are splitted. This technique avoids difficulties with metrics. Up to now, the flux-difference splitting technique only has been used in time-accurate techniques for transient flows and in time-marching techniques for steady flows. Algorithms similar to Roe's were developed by Enquist and Osher [11] and by Lombard et al [12]. In [7,8] it was shown that particularly Roe's scheme can generate steady shocks with only one or even no transition points.

In this paper, it is shown on the example of Cauchy-Riemann equations, that the flux-difference splitting scheme can be used to generate discrete equations which are amenable to a solution with relaxation techniques and that these relaxation schemes have smoothing properties so that they can be used in the multigrid technique with the same efficiency as flux-vector splitting techniques. It is also shown that relaxation techniques based on flux-difference splitting can be constructed for one-dimensional Euler equations, leading to solutions with shocks without transition points.

UPWIND DIFFERENCING

Classic relaxation schemes like Jacobi, Gauss-Seidel and Successive over-relaxation are only proven for positive type equations. These are equations of the form :

$$\alpha_{ii} u_i^h - \alpha_{ij} u_j^h = F_i^h \qquad (1)$$

in which u_i^h denotes the discrete solution in node i while the subscript j describes grid points in the vicinity of i, and with :

(1) positive coefficients : $\alpha_{ii} > 0, \quad \alpha_{ij} \geq 0$

(2) dominance of the central node i : $\alpha_{ii} \geq \sum_{j \neq i} \alpha_{ij}$

(3) irreducibility : the system cannot be decoupled into independent sub-systems.

Classic discretisations of scalar elliptic partial differential equations, as for instance the central discretisation of the Laplace equation, generate difference equations of positive type.

However, it is clear that ellipticity of the partial differential equation is not a necessary condition to achieve difference equations of positive type.

For instance, the scalar steady advection equation :

$$a \frac{\partial u}{\partial x} + b \frac{\partial u}{\partial y} = 0 \qquad (2)$$

leads to a positive type difference equation if upwind differencing is used, i.e. backward differencing for terms in (2) with a positive coefficient and forward differencing for terms with a negative coefficient. Hence, for a>0, b<0 in (2) :

$$a(u_{i,j} - u_{i-1,j}) + b(u_{i,j+1} - u_{i,j}) = 0 \qquad (3)$$

clearly equation (3) can be solved by any standard relaxation scheme.

Furthermore, practical experience shows that positiveness of the difference equations is not a necessary condition for solvability of the system of e-quations by relaxation methods. However, for non-positive equations, non-standard relaxation methods are necessary. A very well known example of this kind is Jameson's rotated difference scheme for transonic potential equations [13]. In supersonic parts of the flow, the potential equation becomes hyperbolic, but due the use of upwind differencing the solution can be obtained by a relaxation method.

It is rather easy to extend the notion of scalar positiveness to vectorial positiveness for systems of first order equations with system matrices with real eigenvalues :

$$A \frac{\partial \xi}{\partial x} + B \frac{\partial \xi}{\partial y} = 0 \ . \tag{4}$$

When A and B have real eigenvalues, it is always possible to split the matrices into a sum of a matrix with positive eigenvalues and a matrix with negative eigenvalues : $\qquad A = A^+ + A^- \qquad\qquad B = B^+ + B^-$

Equation (4) then can be written in split form as :

$$A^+ \frac{\partial^+ \xi}{\partial x} + A^- \frac{\partial^- \xi}{\partial x} + B^+ \frac{\partial^+ \xi}{\partial y} + B^- \frac{\partial^- \xi}{\partial y} = 0 \ . \tag{5}$$

An upwind discretisation of (5) then is obtained when the +terms are discretised by backward differences and the −terms by forward differences :

$$A^+ \left(\xi_{i,j} - \xi_{i-1,j} \right) + A^- \left(\xi_{i+1,j} - \xi_{i,j} \right) + B^+ \left(\xi_{i,j} - \xi_{i,j-1} \right) + B^- \left(\xi_{i,j+1} - \xi_{i,j} \right) = 0$$

or :

$$\left(A^+ + B^+ - A^- - B^- \right) \xi_{i,j} - A^+ \xi_{i-1,j} - \left(-A^- \right) \xi_{i+1,j} - B^+ \xi_{i,j-1} - \left(-B^- \right) \xi_{i,j+1} = 0. \tag{6}$$

Although it is not a general rule, clearly for a large class of system matrices, the coefficient matrix $C = A^+ + B^+ - A^- - B^-$ has positive eigenvalues. In this case, equation (6) is a vector analogue of the scalar equation (1).

It can be called to be of block-positive type since the matrix coefficients then have positive eigenvalues. Although theoretically this might not be completely clear, block-variants of relaxation schemes can be used on block-positive equations.

FLUX-VECTOR SPLITTING FOR CAUCHY-RIEMANN EQUATIONS

The Cauchy-Riemann equations are :

$$\frac{\partial u}{\partial x} + \frac{\partial v}{\partial y} = 0 \qquad\qquad\qquad -\frac{\partial v}{\partial x} + \frac{\partial u}{\partial y} = 0$$

or

$$\begin{pmatrix} 1 & 0 \\ 0 & -1 \end{pmatrix} \frac{\partial}{\partial x} \begin{pmatrix} u \\ v \end{pmatrix} + \begin{pmatrix} 0 & 1 \\ 1 & 0 \end{pmatrix} \frac{\partial}{\partial y} \begin{pmatrix} u \\ v \end{pmatrix} = 0 \ . \tag{7}$$

Equation (7) describes irrotational flow of an inviscid incompressible fluid with velocity components u and v.

The characteristic matrix associated to (7) is :

$$K = \begin{pmatrix} 1 & 0 \\ 0 & -1 \end{pmatrix} k_1 + \begin{pmatrix} 0 & 1 \\ 1 & 0 \end{pmatrix} k_2 = \begin{pmatrix} k_1 & k_2 \\ k_2 & -k_1 \end{pmatrix} .$$

The equation $\mathrm{Det}(K) = -k_1^2 - k_2^2 = 0$ has no non-zero solutions (k_1, k_2). Hence the system of equations (7) is elliptic. It is to be remarked, as discussed in the previous section, that ellipticity is not necessary to arrive at a positive type discretisation.

The eigenvalues of the system matrices are :

for $A = \begin{pmatrix} 1 & 0 \\ 0 & -1 \end{pmatrix}$: $\begin{aligned} \lambda_1 &= 1 \\ \lambda_2 &= -1 \end{aligned}$ hence : $\Lambda_A = \begin{pmatrix} 1 & 0 \\ 0 & -1 \end{pmatrix}$

for B : $\begin{pmatrix} 0 & 1 \\ 1 & 0 \end{pmatrix}$: $\begin{aligned} \lambda_1 &= 1 \\ \lambda_2 &= -1 \end{aligned}$ hence : $\Lambda_B = \begin{pmatrix} 1 & 0 \\ 0 & -1 \end{pmatrix}$

The associated left eigenvector matrices of A and B are

$$E_A = \begin{pmatrix} 1 & 0 \\ 0 & 1 \end{pmatrix} \qquad\qquad E_B = \begin{pmatrix} 1 & 1 \\ 1 & -1 \end{pmatrix}$$

with $E_A \cdot A = \Lambda_A \cdot E_A$ $\qquad E_B \cdot B = \Lambda_B \cdot E_B$.

Since eigenvector matrices can be defined, following Steger and Warming [1] an obvious way of splitting can be obtained by first splitting the eigenvalue matrices :

$$\Lambda_A = \Lambda_A^+ + \Lambda_A^- \qquad\qquad \Lambda_B = \Lambda_B^+ + \Lambda_B^-$$

with, in general, for A (similar for B) :

$$\Lambda_A^+ = \begin{pmatrix} \lambda_{1A}^+ & \\ & \ddots \\ & & \lambda_{nA}^+ \end{pmatrix} \qquad\qquad \Lambda_A^- = \begin{pmatrix} \lambda_{1A}^- & \\ & \ddots \\ & & \lambda_{nA}^- \end{pmatrix}$$

with $\lambda_{iA}^+ = \max(\lambda_{iA}, 0)$ $\qquad \lambda_{iA}^- = \min(\lambda_{iA}, 0)$.

The split matrices are then obtained by :

$$A^+ = E_A^{-1} \Lambda_A^+ E_A \qquad\qquad A^- = E_A^{-1} \Lambda_A^- E_A$$

$$B^+ = E_B^{-1} \Lambda_B^+ E_B \qquad\qquad B^- = E_B^{-1} \Lambda_B^- E_B$$

This gives :

$$A^+ = \begin{pmatrix} 1 & 0 \\ 0 & 0 \end{pmatrix} \quad A^- = \begin{pmatrix} 0 & 0 \\ 0 & -1 \end{pmatrix} \quad B^+ = \begin{pmatrix} .5 & .5 \\ .5 & .5 \end{pmatrix} \quad B^- = \begin{pmatrix} -.5 & .5 \\ .5 & -.5 \end{pmatrix} .$$

The corresponding splitted equations are :

$$\frac{\partial^+ u}{\partial x} + .5 \frac{\partial^+}{\partial y} (u+v) - .5 \frac{\partial^+}{\partial y} (u-v) = 0 \qquad (8)$$

$$- \frac{\partial^- v}{\partial x} + .5 \frac{\partial^+}{\partial y} (u+v) + .5 \frac{\partial^+}{\partial y} (u-v) = 0 \quad . \qquad (9)$$

The coefficient matrix $\ C = A^+ + B^+ - A^- - B^- = \begin{pmatrix} 2 & 0 \\ 0 & 2 \end{pmatrix}$

has only positive eigenvalues.

The corresponding upwind discretisation on a square mesh gives, for interior points :

$$4u_{i,j} = 2u_{i-1,j} + u_{i,j-1} + u_{i,j+1} + v_{i,j-1} - v_{i,j+1} \qquad (10)$$

$$4v_{i,j} = 2v_{i+1,j} + v_{i,j-1} + v_{i,j+1} + u_{i,j-1} - u_{i,j+1} \ . \qquad (11)$$

The boundary conditions to be associated to (8) and (9), on a rectangular region, obviously are :

left boundary (inflow)   : u

right boundary (outflow) : v

upper boundary   : u+v

lower boundary   : u-v

Of course, also a dominant part of this quantities can be specified. In the following numerical examples, the boundary conditions are :

inflow boundary (x = -L) : u = 1

outflow boundary (x = L) : v = 0

upper boundary (y = H)   : v = 0

lower boundary (y = -H) : v = αu

with $\alpha = 0$ for $x \leq -2L/3$ and $x \geq 2L/3$

$$\alpha = (3x/8L)\left((3x/2L)^2 - 1\right) \quad \text{for} \quad -2L/3 \leq x \leq 2L/3 \ .$$

At inflow, the v-equation (11) and at outflow, the u-equation (10) can be used. At the upper boundary the equations (8) and (9) are to be combined in order to eliminate $\partial^-/\partial y$:

$$\frac{\partial^+ u}{\partial x} - \frac{\partial^- v}{\partial x} + \frac{\partial^+}{\partial y} (u+v) = 0 \ .$$

This gives, with $v_{i,j} = v_{i+1,j} = 0$:

$$2u_{i,j} = u_{i-1,j} + u_{i,j-1} + v_{i,j-1} \ .$$

The same equation can be used at the right upper corner with $v_{i,j-1} = 0$.
At the lower boundary, the equations (8) and (9) are to be combined in order to eliminate $\partial^+/\partial y$:

26

$$\frac{\partial^+ u}{\partial x} + \frac{\partial^- v}{\partial x} - \frac{\partial^-}{\partial y} (u-v) = 0 \ .$$

This gives, with $v_{i,j} = \alpha u_{i,j}$

$$2\left(1 - \alpha_{i,j}\right) u_{i,j} = u_{i-1,j} + u_{i,j+1} - v_{i+1,j} - v_{i,j+1} \ .$$

At the right lower corner, taking into account $\alpha_{i,j} = 0$ and $v_{i,j+1} = 0$, this becomes :

$$2u_{i,j} = u_{i-1,j} + u_{i,j+1} \ .$$

The resulting set of equations can be solved by block-Jacobi, block-successive relaxation, both in lexicographic (LX) ordering and Red-Black (RB) ordering. From a practical point of view the stability of these schemes can simply be verified by the construction, through Taylor expansion, of the equivalent time dependent equations, associated to the relaxation schemes. It is found that the equivalent equations are hyperbolic with respect to the pseudo-time for relaxation factors lower than some upper values.

NUMERICAL EXAMPLES

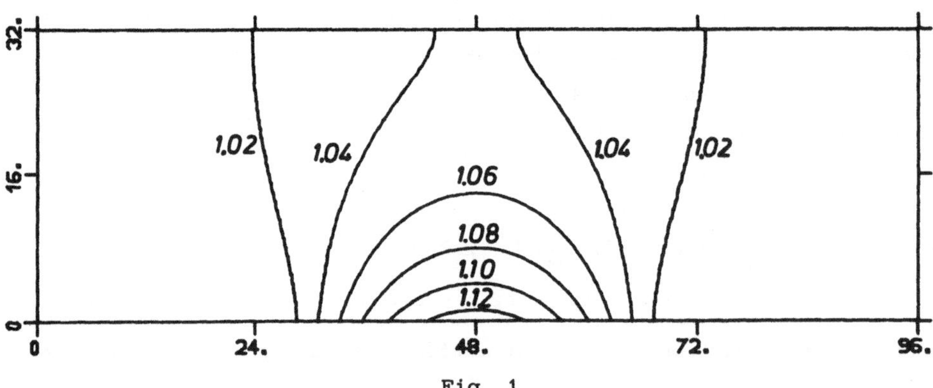

Fig. 1

Figure 1 shows the lines of constant total velocity for the solution obtained by relaxation of the Cauchy-Riemann equations, discretised by flux-vector splitting on a 96 x 32 elements mesh. Both red-black ordering and lexicographic ordering were used. It was found that a symmetric lexicographic scheme (SLX) performs the best on a single grid. In the SLX-scheme the nodes were relaxed by alternating sweeps from the lower left point to the upper right point, first varying the row index and sweeps from the upper right point to the lower left point, also first varying the row index. For both the RB-scheme and the SLX-scheme, optimum acceleration was reached for

ω = .95. Both schemes were used in multigrid form, using four grids : 96x32, 48x16, 24x8 and 12x4, V-cycling, full weighting as restriction operator and bilinear interpolation as prolongation operator. In both RB and SLX-schemes the optimal smoothing experimentally was found to be reached for the relaxation factor ω = .75. The optimum configuration in both schemes was found to be :

RB and LX	ν_1	ν_2
h	1	0
2h	2	1
4h	3	2
8h	8	

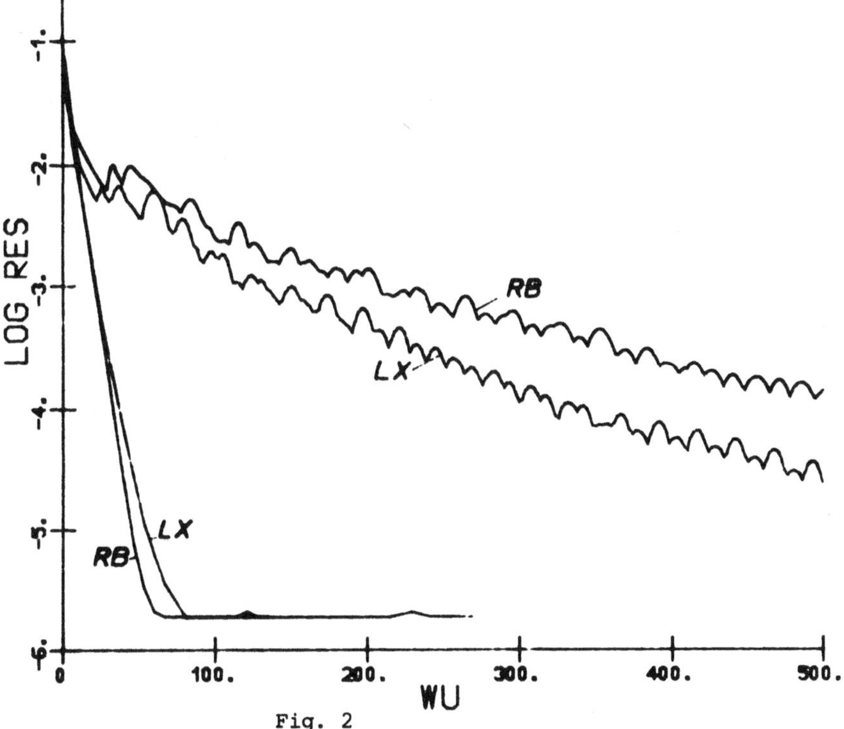

Fig. 2

Figure 2 shows the convergence results. The residual shown is the maximum residual. The work units in the multigrid scheme were determined as :
one relaxation sweep = 1 wu on grid h, 1/4 wu on grid 2h, ...
one residual calculation + the associated grid transfer = 1 wu on grid h, 1/4 wu on grid 2h, ...

According to this rule, the resulting number of work units per cycle is :

$$(\nu_{1,h} + \nu_{2,h} + 1) + \frac{1}{4}(\nu_{1,2h} + \nu_{2,2h} + 1) + \frac{1}{16}(\nu_{1,4h} + \nu_{2,4h} + 1) + \frac{1}{64}(\nu_{1,8h} + \nu_{2,8h}) .$$

Actual measurement of computing times shows that this formula underestimates the work by 3 à 5 % for all parameter combinations.

In figure 2, it is seen that the RB-scheme performs better than the LX-scheme in multigrid version. The convergence factor for the RB-scheme is :

$$\rho = \frac{\text{maximum residual (wu = I)}}{\text{maximum residual (wu = I-1)}} \cong .700 .$$

This is only slightly less performing than a typical multigrid algorithm on a scalar equation (MGOO : $\rho \cong .5$ to $.6$).

FLUX-DIFFERENCE SPLITTING VERSUS FLUX-VECTOR SPLITTING

The approach used in the preceeding paragraph, introduced by Steger and Warming has been called flux-vector splitting. It was devised for Euler equations. These take the form :

$$\frac{\partial F}{\partial x} + \frac{\partial G}{\partial y} = 0 \qquad F = \begin{pmatrix} \rho u \\ \rho uu+p \\ \rho uv \\ \rho Hu \end{pmatrix} \qquad G = \begin{pmatrix} \rho v \\ \rho uv \\ \rho vv+p \\ \rho Hv \end{pmatrix} . \qquad (12)$$

ρ is density, u and v are Cartesian velocity components, $H = E + \frac{p}{\rho}$ is total enthalpy, p is pressure, $E = \frac{1}{\gamma-1} \frac{p}{\rho} + \frac{1}{2} u^2 + \frac{1}{2} v^2$ is total energy and γ is adiabatic constant (air $\gamma = 1.4$). The flux vectors F and G have the remarkable property that they are homogeneous with respect to the variables $\xi_f^T = (f_1, f_2, f_3, f_4) = (\rho, \rho u, \rho v, \rho E)$.

It is easily verified that :

$$F = \begin{pmatrix} f_2 \\ \dfrac{3-\gamma}{2} \dfrac{f_2^2}{f_1} - \dfrac{\gamma-1}{2} \dfrac{f_3^2}{f_1} + (\gamma-1) f_4 \\ \dfrac{f_2 f_3}{f_1} \\ -\dfrac{\gamma-1}{2} \dfrac{f_2^3}{f_1^2} - \dfrac{\gamma-1}{2} \dfrac{f_2 f_3^2}{f_1^2} + \gamma \dfrac{f_2 f_4}{f_1} \end{pmatrix} \qquad G = \begin{pmatrix} f_3 \\ \dfrac{f_2 f_3}{f_1} \\ -\dfrac{\gamma-1}{2} \dfrac{f_2^2}{f_1} + \dfrac{3-\gamma}{2} \dfrac{f_3^2}{f_1} + (\gamma-1) f_4 \\ -\dfrac{\gamma-1}{2} \dfrac{f_2^2 f_3}{f_1^2} - \dfrac{\gamma-1}{2} \dfrac{f_3^3}{f_1^2} + \gamma \dfrac{f_3 f_4}{f_1} \end{pmatrix}$$

Obviously : $F(\alpha\xi_f) = \alpha F(\xi_f)$ $\qquad\qquad$ $G(\alpha\xi_f) = \alpha G(\xi_f)$

Hence by defining the Jacobians : $A = \dfrac{\partial F}{\partial \xi_f}$ $\qquad B = \dfrac{\partial G}{\partial \xi_f}$

(12) can be written as : $A \dfrac{\partial \xi_f}{\partial x} + B \dfrac{\partial \xi_f}{\partial y} = 0$.   (13)

Equation (13) is not a linearisation of (12) but is completely equivalent with it. It was demonstrated by Jespersen [6] that the flux vector splitting approach for Euler equations leads to discrete equations which can be solved in the same way as was done in the preceeding section for Cauchy-Riemann equations. Since in (13) the Jacobians vary from point to point, the resulting discretisation is not telescoping, i.e. when equations in neighbouring points are added, the corresponding forward and backward terms do not cancel identically. Therefore the discrete equations do not express the conservation property inherent in the corresponding differential equations. As a consequence, in transonic applications errors are found in sonic regions and around shocks. Complete conservation can be reached by

bringing the flux-vector splitting in control volume formulation according to figure 3. On the cell surface $S_{i+1/2}$, fluxes are split into a backward and a forward part. The backward part is expressed by the state variables in

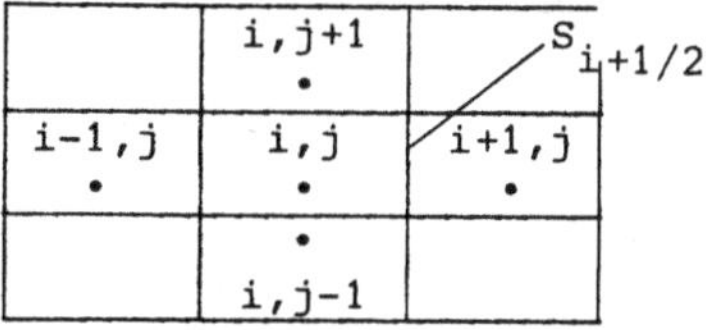

Fig. 3

node (i,j) and the forward part by the state variables in the node (i+1,j). The other surfaces are treated similarly. There is an obvious difficulty in defining the split parts of flux-vector components associated to an eigenvalue which changes sign when calculated with the state variables in the nodes on both sides of the cell surface. Correct splitting only can be done by defining transition cells leading to two transition points in a shock [8].

Difficulties of this kind do not occur when (12) is discretised on a finite volume and when the differences of fluxes between opposite surfaces are splitted [9]. This is possible due to the homogeneity of degree two of the flux-vectors with respect to the variables

$$\xi_g^T = \left(g_1, g_2, g_3, g_4\right) = \left(\sqrt{\rho},\ \sqrt{\rho}\,u,\ \sqrt{\rho}\,v,\ \sqrt{\rho}\,H\right) .$$

It is easy to verify that :

$$F = \begin{pmatrix} g_1 g_2 \\ \dfrac{\gamma+1}{2\gamma}\, g_2^2 - \dfrac{\gamma-1}{2\gamma}\, g_3^2 + \dfrac{\gamma-1}{\gamma}\, g_1 g_4 \\ g_2 g_3 \\ g_2 g_4 \end{pmatrix} \qquad G = \begin{pmatrix} g_1 g_3 \\ g_2 g_3 \\ -\dfrac{\gamma-1}{2\gamma}\, g_2^2 + \dfrac{\gamma+1}{2\gamma}\, g_3^2 + \dfrac{\gamma-1}{\gamma}\, g_1 g_4 \\ g_3 g_4 \end{pmatrix} .$$

Clearly
$$F(\alpha \xi_g) = \alpha^2 F(\xi_g) \qquad\qquad G(\alpha \xi_g) = \alpha^2 G(\xi_g)$$

As a consequence, differences of fluxes can be written exactly as :
$$\Delta F = \overline{A} \, \Delta \xi_g \qquad\qquad \Delta G = \overline{B} \, \Delta \xi_g$$

in which $\overline{A}$ and $\overline{B}$ are Jacobians defined on mean values. For example, on the first component of F, the splitting can be done according to :

$$\mathop{\Delta}_{i,j} F_1 = \mathop{\Delta}_{i,j} g_1 g_2 = (g_1 g_2)_i - (g_1 g_2)_j = \frac{g_{1,i} + g_{1,j}}{2}(g_{2,i} - g_{2,j}) + \frac{g_{2,i} + g_{2,j}}{2}(g_{1,i} - g_{1,j}). \quad (14)$$

FLUX-DIFFERENCE SPLITTING FOR ONE-DIMENSIONAL EULER EQUATIONS .

In order to show how flux-difference splitting can be used on Euler equations, we consider steady transonic flow in a one-dimensional nozzle of form :

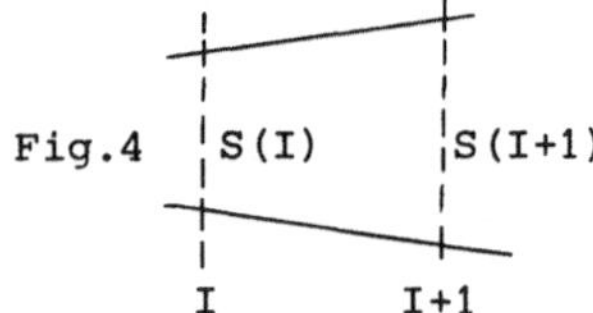

$$S(x) = .9 + .1 \left(2(3x/2L)^2 - (3x/2L)^4\right) \quad \text{for} \quad -\frac{2L}{3} \le x \le \frac{2L}{3}$$

$$S(x) = 1 \quad \text{for} \quad -L \le x \le -\frac{2L}{3} \quad \text{and} \quad \frac{2L}{3} < x \le L .$$

On the segment shown in figure 4, the discrete equations are :

Mass : $\qquad\qquad\qquad$ RUS(I+1) − RUS(I) = 0 $\qquad\qquad\qquad\qquad\qquad$ (15)

Momentum : RUUS(I+1) − RUUS(I) + .5$\big($S(I+1)+S(I)$\big)\big($P(I+1)−P(I)$\big)$ = 0 $\qquad$ (16)

Energy : $\qquad\qquad\qquad$ RHUS(I+1) − RHUS(I) = 0 $\qquad\qquad\qquad\qquad\qquad$ (17)

in which, in the discrete system, R denotes density, U velocity, P pressure, H total enthalpy and S section. The force term in the momentum equation is obtained by prescribing that pressure and section vary lineary between nodes.

The splitting of Roe is not applicable since the system (15-17) is not in strong conservation form. Another splitting, with similar properties, can be used, which, however, cannot be extended to two dimensions.

Using the splitting rule (14), (15) becomes :

$$\frac{RU(I+1)+RU(I)}{2}\big(S(I+1)-S(I)\big) + \frac{S(I+1)+S(I)}{2}\big(RU(I+1)-RU(I)\big)$$

and $\quad \dfrac{U(I+1)+U(I)}{2}\big(R(I+1)-R(I)\big) + \dfrac{R(I+1)+R(I)}{2}\big(U(I+1)-U(I)\big)$

$$+ \frac{RU(I+1)+RU(I)}{2}\big(S(I+1)-S(I)\big) = 0$$

or : $\qquad\qquad\qquad \overline{U}\,\Delta R + \overline{R}\,\Delta U = -\frac{\overline{RU}}{S}\,\Delta S \qquad\qquad\qquad$ (18)

Combination of (15) and (16) gives :

$$\frac{RUS(I+1)+RUS(I)}{2}\Big(U(I+1)-U(I)\Big) + \frac{S(I+1)+S(I)}{2}\Big(P(I+1)-P(I)\Big)$$

or
$$\overline{U}\,\Delta U + \frac{1}{\overline{\overline{R}}}\,\Delta P = 0 \tag{19}$$

with :
$$\overline{\overline{R}} = \frac{\Big(RUS(I+1)+RUS(I)\Big)/2}{\Big(U(I+1)+U(I)\Big)/2\;\Big(S(I+1)+S(I)\Big)/2}\;.$$

Substitution of $RH = \frac{\gamma}{\gamma-1}P + \frac{1}{2}RUU$ into (17) gives :

$$\frac{1}{2}\Big(RUUUS(I+1) - RUUUS(I)\Big) + \frac{\gamma}{\gamma-1}\Big(PUS(I+1) - PUS(I)\Big) = 0$$

Combination with (15) gives :

$$\frac{RUS(I+1)+RUS(I)}{2}\,\frac{U(I+1)+U(I)}{2}\Big(U(I+1)-U(I)\Big) + \frac{\gamma}{\gamma-1}\,\frac{PU(I+1)+PU(I)}{2}\Big(S(I+1)-S(I)\Big)$$

$$+ \frac{\gamma}{\gamma-1}\,\frac{S(I+1)+S(I)}{2}\Big(PU(I+1)-PU(I)\Big) = 0$$

Using (19) :
$$\gamma\,\overline{P}\,\Delta U + \overline{U}\,\Delta P = -\gamma\,\frac{\overline{PU}}{\overline{S}}\,\Delta S\;. \tag{20}$$

The equations (18)(19)(20) form a system :

$$\begin{pmatrix} \overline{U} & \overline{R} & 0 \\ 0 & \overline{U} & 1/\overline{\overline{R}} \\ 0 & \gamma\overline{P} & \overline{U} \end{pmatrix} \begin{pmatrix} \Delta R \\ \Delta U \\ \Delta P \end{pmatrix} = \begin{pmatrix} -\overline{RU}/\overline{S}\,\Delta S \\ 0 \\ -\gamma\overline{PU}/\overline{S}\,\Delta S \end{pmatrix}\;. \tag{21}$$

The system matrix has the eigenvalues : $\overline{U}$, $\overline{U}+\overline{C}$, $\overline{U}-\overline{C}$ with $\overline{C} = \sqrt{\gamma\overline{P}/\overline{\overline{R}}}$

The corresponding left eigenvector matrix is :

$$E = \begin{pmatrix} 1 & 0 & -\overline{R}/\overline{P} \\ 0 & 1 & 1/\sqrt{\gamma\overline{P}\;\overline{\overline{R}}} \\ 0 & 1 & -1/\sqrt{\gamma\overline{P}\;\overline{\overline{R}}} \end{pmatrix}\;. \tag{22}$$

By premultiplying (21) by (22), the equations become, with $\overline{M} = \overline{U}/\overline{C}$:

$$\frac{\Delta R}{\overline{R}} - \frac{1}{\gamma}\frac{\Delta P}{\overline{P}} = \Big(\frac{\overline{PU}}{\overline{P}\,\overline{U}} - \frac{\overline{RU}}{\overline{R}\,\overline{U}}\Big)\frac{\Delta S}{\overline{S}} \tag{23}$$

$$(\overline{M}+1)\frac{\Delta U}{\overline{U}} + \frac{1}{\gamma}\Big(\frac{1+\overline{M}}{\overline{M}}\Big)\frac{\Delta P}{\overline{P}} = \frac{\overline{PU}}{\overline{P}\,\overline{U}}\frac{\Delta S}{\overline{S}} \tag{24}$$

$$(\overline{M}-1)\frac{\Delta U}{\overline{U}} + \frac{1}{\gamma}\Big(\frac{1-\overline{M}}{\overline{M}}\Big)\frac{\Delta P}{\overline{P}} = -\frac{\overline{PU}}{\overline{P}\,\overline{U}}\frac{\Delta S}{\overline{S}}\;. \tag{25}$$

In subsonic flow, equations (23) and (24) are associated to a positive eigenvalue, while equation (25) is associated to a negative eigenvalue. Hence, in this case, in the interval (I,I+1), (23) and (24) are to be used to express values at node I+1 in function of values at node I, while (25) is to be used to express values at node I in function of values at node I+1. Hence the nodal equations at node I consist of (23) and (24) for the interval (I-1,I) and (25) for the interval (I,I+1). To close the system of equations, two boundary conditions are to be given at inflow and one boundary condition at outflow. It is clear that these conditions are to be stagnation enthalpy and stagnation pressure at inflow and pressure at ouflow.

When the flow is supersonic in some interval, all equations (23-25) have to express values at node I+1 in function of values at node I. There is an obvious difficulty for a node adjacent to an interval which is subsonic and an interval which is supersonic. According to the above rule, either two equations or four equations are assigned to this node. This difficulty can be avoided by writing the system (23-25) as an assemblage of two systems :

$$\begin{pmatrix} 1 & 0 & -\overline{R}/\gamma\overline{P} \\ 0 & (1+\overline{M}) & \dfrac{(1+\overline{M})\overline{U}}{\gamma\overline{M}\ \overline{P}} \\ 0 & \dfrac{\overline{M}-M^*}{M^*} & -\dfrac{(\overline{M}-M^*)\overline{U}}{\gamma\overline{MM^*}\ \overline{P}} \end{pmatrix}_{I-1,I} \begin{pmatrix} R(I)-R(I-1) \\ U(I)-U(I-1) \\ P(I)-P(I-1) \end{pmatrix} = \begin{pmatrix} (\dfrac{\overline{PU}}{\overline{P}}\ \overline{R}-\overline{RU})\ \dfrac{\Delta S}{\overline{S}\ \overline{U}} \\ -\dfrac{\overline{PU}}{\overline{P}}\ \dfrac{\Delta S}{\overline{S}} \\ \dfrac{\overline{M}-M^*}{2\overline{M}-M^*-\overline{MM^*}}\ \dfrac{\overline{PU}}{\overline{P}}\ \dfrac{\Delta S}{\overline{S}} \end{pmatrix}_{I-1,I} \qquad (26)$$

$$\begin{pmatrix} 0 & 0 & 0 \\ 0 & 0 & 0 \\ 0 & -\dfrac{M(1-M^*)}{M^*} & \dfrac{\overline{M}(1-M^*)}{\gamma\overline{MM^*}}\ \dfrac{\overline{U}}{\overline{P}} \end{pmatrix}_{I,I+1} \begin{pmatrix} R(I+1)-R(I) \\ U(I+1)-U(I) \\ P(I+1)-P(I) \end{pmatrix} \begin{pmatrix} 0 \\ 0 \\ \dfrac{\overline{M}-M^*\overline{M}}{2\overline{M}-M^*-\overline{MM^*}}\ \dfrac{\overline{PU}}{\overline{P}}\ \dfrac{\Delta S}{\overline{S}} \end{pmatrix}_{I,I+1} . \qquad (27)$$

The assemblage gives the correct results in subsonic flow for $M^* = \overline{M}$ and in supersonic flow for $M^* = 1$. In order to avoid singularities in sonic zones, in practice one can choose $M^* = M$ for $M < 1-\varepsilon$ and $M^* = 1-\varepsilon$ for $M^* \geq 1-\varepsilon$, in which ε is as close as possible to 1.

Figure 5 shows the result obtained for a discretisation with 96 elements, back pressure equal to .72 of the stagnation inlet pressure, for air ($\gamma = 1.4$), using symmetric Gauss-Seidel relaxation ($\omega = 1$) and $\varepsilon = .007$. The shock profile is obtained without transition points and except for a small disturbance in the sonic zone, the computed and exact solutions coincide. Results obtained for lower values of ε show a larger discrepancy in the sonic region while results for larger values of ε usually have one transition point. Of course for very large values of ε, the consistency error due to the use of ε becomes large, resulting in a discrepancy between calculated and exact solutions behind the shock.

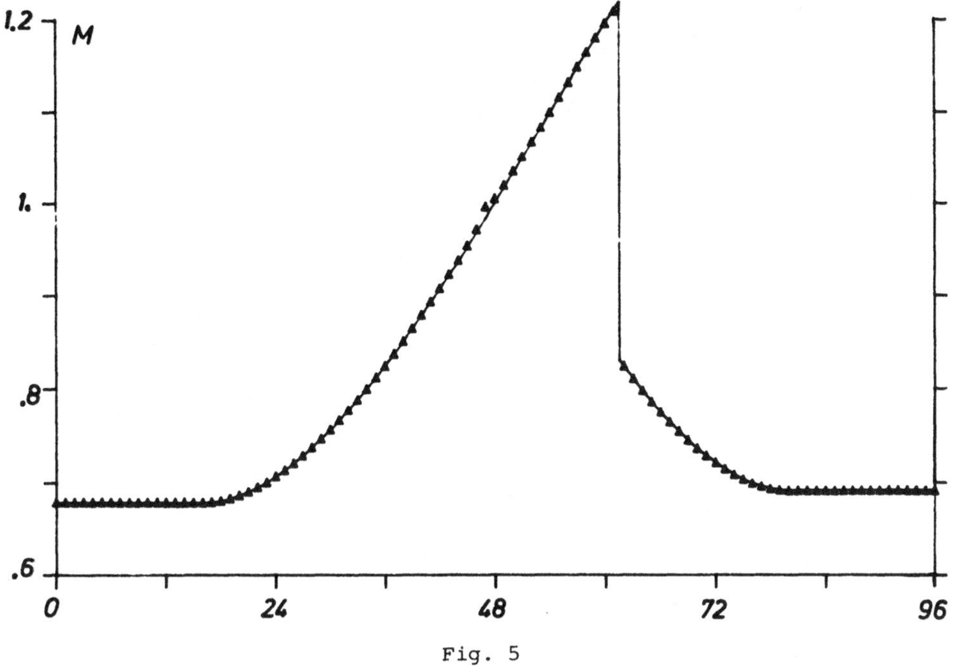

Fig. 5

TRUTH MATRIX FORMULATION

The splitting described in the previous section can be done in a more systematic way, using the truth matrix approach introduced by Lombard et al
[12]. Associated to the positive and negative parts of the eigenvalue matrix,
truth matrices can be defined. These are diagonal matrices with elements 1
corresponding to non-zero elements in the splitted eigenvalue matrices.
For example :

$$\Lambda^+ = \begin{pmatrix} \lambda_1^+ & & \\ & \lambda_2^+ & \\ & & 0 \end{pmatrix} \rightarrow D^+ = \begin{pmatrix} 1 & & \\ & 1 & \\ & & 0 \end{pmatrix} \quad \Lambda^- = \begin{pmatrix} 0 & & \\ & 0 & \\ & & \lambda_3^- \end{pmatrix} \rightarrow D^- = \begin{pmatrix} 0 & & \\ & 0 & \\ & & 1 \end{pmatrix} .$$

Premultiplicator matrices then can be constructed as :

$$M_A^+ = E^{-1} D^+ E \quad \text{and} \quad M_A^- = E^{-1} D^- E \quad \rightarrow \quad M_A^+.A = A^+ \quad \text{and} \quad M_A^-.A = A^-$$

It is easy to verify that an assemblage equivalent to (26) and (27) is obtained by premultiplication of the system (23-25) by M_A^+ in the interval
(I-1,I) and by M_A^- in the interval (I,I+1). A perturbation in the sonic region, avoiding the singularity of the system, then can be obtained by adding a perturbation matrix A_ε to the system matrix A^+ in (I-1,I) and by
substracting A_ε from the system matrix A^- in (I,I+1), with :

$$A = E^{-1} \begin{pmatrix} 0 & & \\ & 0 & \\ & & \varepsilon^* \overline{c} \end{pmatrix} E \qquad \begin{aligned} \varepsilon^* &= 0 && \text{for} && M \leq 1-\varepsilon \\ \varepsilon^* &= M+\varepsilon-1 && \text{for} && 1-\varepsilon \leq M \leq 1 \\ \varepsilon^* &= \varepsilon && \text{for} && M \geq 1 \end{aligned}$$

FLUX-DIFFERENCE SPLITTING FOR CAUCHY-RIEMANN EQUATIONS

In order to illustrate the flux-difference approach in two dimensions, we consider the discretisation of Cauchy-Riemann equations on a square grid, using finite volumes with bilinear interpolation. On the element shown in figure 6, the discretisation of (7) gives :

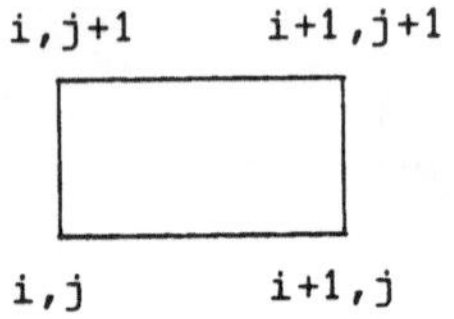

$$\frac{1}{2}(u_{i+1,j+1} + u_{i+1,j}) - \frac{1}{2}(u_{i,j+1} + u_{i,j}) + \frac{1}{2}(v_{i+1,j+1} + v_{i,j+1}) - \frac{1}{2}(v_{i+1,j} + v_{i,j}) = 0$$

$$\frac{1}{2}(v_{i,j+1} + v_{i,j}) - \frac{1}{2}(v_{i+1,j+1} + v_{i+1,j}) + \frac{1}{2}(u_{i+1,j+1} + u_{i,j+1}) - \frac{1}{2}(u_{i+1,j} + u_{i,j}) = 0 .$$

This can be written as :

$$A \underset{x}{\Delta} \xi + B \underset{y}{\Delta} \xi = 0 \tag{28}$$

in which $\underset{x}{\Delta}$ and $\underset{y}{\Delta}$ correspond to differences of mean values on opposite sides of the finite volume.

According to the previous section, the (constant) premultiplicator matrices for Cauchy-Riemann equations are :

$$M_A^+ = \begin{pmatrix} 1 & 0 \\ 0 & 0 \end{pmatrix} \qquad M_A^- = \begin{pmatrix} 0 & 0 \\ 0 & 1 \end{pmatrix} \qquad M_B^+ = \begin{pmatrix} .5 & .5 \\ .5 & .5 \end{pmatrix} \qquad M_B^- = \begin{pmatrix} .5 & -.5 \\ -.5 & .5 \end{pmatrix} .$$

The premultiplied equations are :

$$A^+ \text{eq.} \; : \; M_A^+ A \underset{x}{\Delta} \xi + M_A^+ B \underset{y}{\Delta} \xi = 0 \tag{29}$$

$$\begin{pmatrix} 1 & 0 \\ 0 & 0 \end{pmatrix} \qquad \begin{pmatrix} 0 & 1 \\ 0 & 0 \end{pmatrix}$$

$$A^- \text{eq.} \; : \; M_A^- A \underset{x}{\Delta} \xi + M_A^- B \underset{y}{\Delta} \xi = 0 \tag{30}$$

$$\begin{pmatrix} 0 & 0 \\ 0 & -1 \end{pmatrix} \qquad \begin{pmatrix} 0 & 0 \\ 1 & 0 \end{pmatrix}$$

$$B^+ \text{eq.} \; : \; M_B^+ A \underset{x}{\Delta} \xi + M_B^+ B \underset{y}{\Delta} \xi = 0 \tag{31}$$

$$\begin{pmatrix} .5 & -.5 \\ .5 & -.5 \end{pmatrix} \qquad \begin{pmatrix} .5 & .5 \\ .5 & .5 \end{pmatrix}$$

$$B^- \text{eq.} : M_B^- A \underset{x}{\Delta} \xi + M_B^- B \underset{y}{\Delta} \xi = 0 \tag{32}$$

$$\begin{pmatrix} .5 & .5 \\ -.5 & -.5 \end{pmatrix} \begin{pmatrix} -.5 & .5 \\ .5 & -.5 \end{pmatrix} .$$

A two-dimensional analogue of the one-dimensional flux-difference splitting
is obtained if the equations (29-32) are assembled according to the pattern :

A^+eq + B^-eq	A^-eq + B^-eq
A^+eq + B^+eq	A^-eq + B^+eq

The resulting system of equations is not block-positive and hence it is not
clear a priori that it can be solved by a relaxation method. Experimentally
it is found that only very few relaxation patterns are stable. An example
of a stable scheme is the four-coloured scheme :

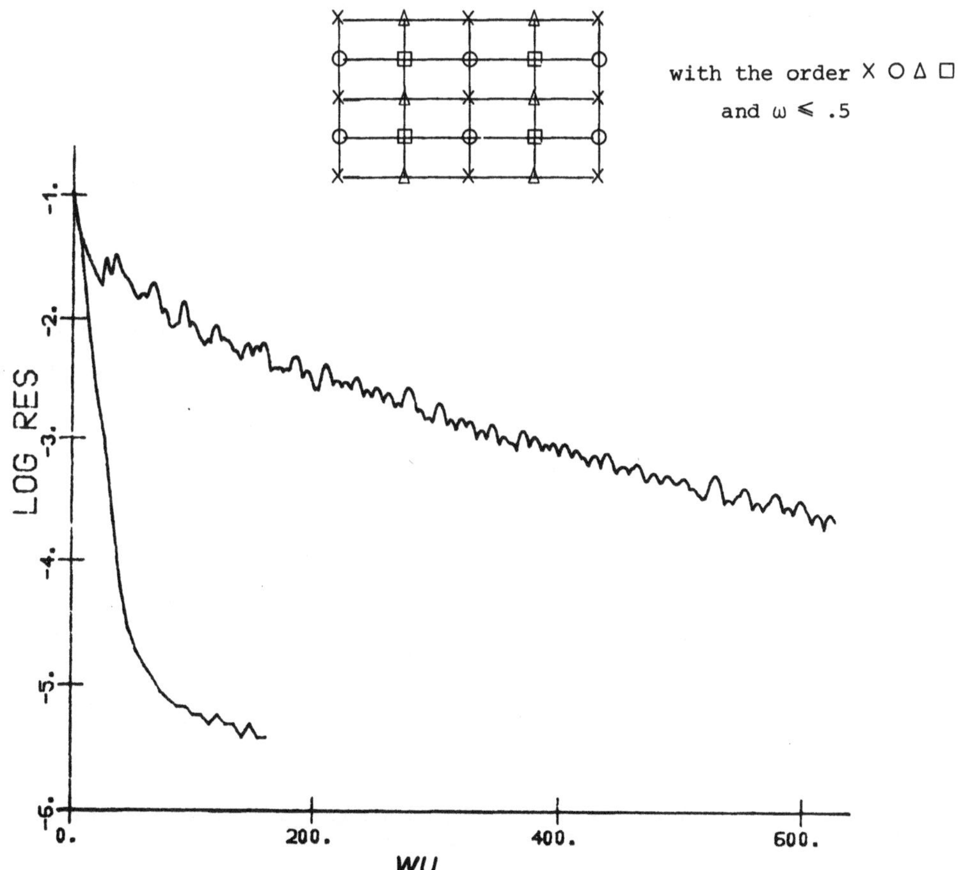

with the order X O Δ □
and $\omega \leqslant .5$

Fig. 7

Figure 7 shows the convergence history for this scheme both in single grid and in multigrid on the problem of figure 1, for $\omega = .475$. The optimum cycle is found to be a saw-tooth cycle (i.e. $\nu_2 = 0$) with parameters :

	h	2h	4h	8h
$\nu_1 =$	4	4	4	10

The work units are calculated with the previously derived formula. By comparison with figure 2, it is seen that the flux-difference splitting method is slightly less performing than the flux-vector splitting method.

EXTENSION TO EULER EQUATIONS

It is clear that Euler equations can be treated in the same way as Cauchy-Riemann equations, when the splitting of Roe is used. Of course, the system matrices vary from element to element and hence a full approximation scheme (FAS) is to be used. Preliminary results obtained up to now for subsonic flows are encouraging but some difficulties still are left in transonic flows.

ACKNOWLEDGMENT

Part of this research was performed while the author was in residence at the NASA Ames Research Center on a grant of the Belgian National Science Foundation (NFWO).

REFERENCES

1   Steger, J.L., and Warming, R.F. : Flux Vector Splitting of the Inviscid Gasdynamic Equations with Application to Finite-Difference Methods; J. Comp. Phys., Vol. 40, 1981, p. 263-293.

2   Courant, R., Isaacson, E. and Rees, M. : On the Solution of Nonlinear Hyperbolic Differential Equations by Finite Differences; Comm. Pure Appl. Math., Vol. 5, 1952, p. 243-255.

3   Reklis, R.D., and Thomas, P.D. : A Shock Capturing Algorithm for the Navier-Stokes Equations; AIAA paper 81-1021, 1981.

4   Buning P.G., and Steger, J.L. : Solution of the Two-Dimensional Euler
     Equations with Generalized Coordinate Transformation using Flux Vector
     Splitting; AIAA paper 82-0971, 1982.

5   Deese, J.E. : Numerical Experiments with the Split-Flux-Vector Form
     of the Euler Equations. AIAA paper 83-0122, 1983.

6   Jespersen, D.C. : A Multigrid Method for the Euler Equations;
     AIAA paper 83-0124, 1983

7   Osher S., and Chakravarthy S. : Upwind Schemes and Boundary Condi-
     tions with Applications to Euler Equations in General Geometries;
     J. Comp. Phys., vol. 50, 1983, p. 447-481.

8   Van Leer, B. :
     SIAM J. Sci. Stat. Comp. (to appear in March 1984)

9   Roe, P.L. : Approximate Riemann Solvers, Parameter Vectors and Diffe-
     rence Schemes; J. Comp. Phys., Vol. 43, 1981, p. 357-372

10   Godunov, S.K. : A Finite-Difference Method for the Numerical Computa-
     tion of Discontinuous Solutions of the Equations of Fluid Dynamics;
     Matematicheskii Sbornik, Vol. 43, 1959, p. 271-290.

11   Enquist, B., and Osher, S. : One Sided Difference Equations for Non-
     linear Conservation Laws; Math. Comp. Vol. 36, 1981, p. 321-352.

12   Lombard, C.K., Oliger, J., and Yang, J.Y. : A Natural Conservative
     Flux Difference Splitting for the Hyperbolic Systems of Gasdynamics;
     AIAA paper 82-0976, 1982.

13   Jameson A. : Iterative Solution of Transonic Flows over Airfoils and
     Wings, Including Flows at Mach 1; Comm. Pure Appl. Math., Vol. 27,
     1974

MULTI-GRID SCHEMES FOR INCOMPRESSIBLE FLOWS

L. Fuchs
Department of Gasdynamics
The Royal Institute of Technology
S-100 44 Stockholm, Sweden

SUMMARY

A new finite-difference scheme for the incompressible Navier-Stokes equations in two space dimensions is formulated. This scheme uses the velocity vector and the pressure gradients as dependent variables. A Multi-Grid scheme for the system of equations is proposed. The algebraic convergence rate of the new (PGA-MG) scheme is compared to the convergence of the (DGS-MG) scheme of Brandt and Dinar [1]. As test problems we use either a smooth test case, the flow in a driven cavity or the flow in a separator model. These problems represent flows with decreasing order of smoothness. It was found that the new PGA-MG scheme has a good convergence rate even for those Reynolds numbers for which the DGS-MG scheme is unstable.

INTRODUCTION

We consider the numerical solution of the incompressible, steady Navier-Stokes equations in a two-dimensional planar geometry. We use here the primitive variable form of the governing equations since we intend to use the developed method for three-dimensional cases as well. A Multi-Grid (MG) method for the solution of the two-dimensional problem was proposed by Brandt and Dinar [1], where they introduced the Distributive Gauss-Seidel (DGS) scheme for the relaxations. A variant of that method was used for some two- and three-dimensional problems [2,3]. In some of the computations it was observed that with increasing Reynolds number (Re) the convergence of the method became slower. For even larger Re the numerical procedure diverged. The difficulties with the DGS scheme, were attributed to some of the assumptions which are associated with the method. In a previous paper [4], some early attempts to overcome the high Re difficulty, were reported. In this work we describe a new scheme which turned out to be more robust compared to other methods.

The new scheme uses the components of the velocity vector and the pressure gradients as dependent variables. A relaxation scheme for updating the pressure gradients is developed. This updating of the pressure gradients is called here as the Pressure Gradient Averaging (PGA) scheme. The PGA and the corresponding DGS schemes have been incorporated into a MG solver. More details about these methods are given in the following.

The two methods were tested on three problems, with different levels of smoothness. As for other problems, it was found that the DGS-MG scheme diverges   for high Re (the critical Re depends mainly on the problem and in a minor way also on the smoothing scheme which is used for the momentum equations). The PGA-MG scheme for the corresponding problems gave converged solutions. For smooth cases the convergence factors of both the DGS and the PGA schemes are comparable. But,however, since the former scheme needs less computation it may be preferable.

THE DISCRETE PROBLEM

The steady state incompressible NS equations in two space dimensions are given by:

$$\nabla^2 u - Re(u\, u_x + v\, u_y) - p_x = 0 \tag{1.a}$$

$$\nabla^2 v - Re(u\, v_x + v\, v_y) - p_y = 0 \tag{1.b}$$

$$u_x + v_y = 0 . \tag{1.c}$$

We use no-slip boundary conditions on all solid boundaries.

Equations (1) are discretized on a staggered grid. All the derivatives are approximated by central differences, except the first derivatives of the convective terms which are approximated by upwind differences. The finite-difference equations are written as

$$Qu - \partial_x p = 0 \tag{2.a}$$

$$Qv - \partial_y p = 0 \tag{2.b}$$

$$\partial_x u + \partial_y v = 0 \tag{2.c}$$

where $Q = \partial_x^2 + \partial_y^2 - Re(u\overset{+}{\partial}_x + v\overset{+}{\partial}_y)$ whereas ∂_x, ∂_y and ∂_x^2, ∂_y^2 are the central difference approximations to the first and the second derivatives, respectively. $\overset{+}{\partial}_x$ and $\overset{+}{\partial}_y$ denote the upwind finite-difference approximations to the first derivatives.

For the PGA scheme we use (p_x) and (p_y) as dependent variables. The system of differential equations (1) is completed then by the identity

$$(p_x)_y = (p_y)_x \ . \tag{1.d}$$

This equation is also approximated on the same staggered grid, by central finite-differences. Using the same notations as in equations (2), the dif-ference approximation to (1.d) is:

$$\partial_y(p_x) = \partial_x(p_y) \ . \tag{2.d}$$

The dependent variables are defined on the sides of each computational cell, whereas equation (2.a) is satisfied in the center and (2.d) at the corner of each computational cell (see Figure 1).

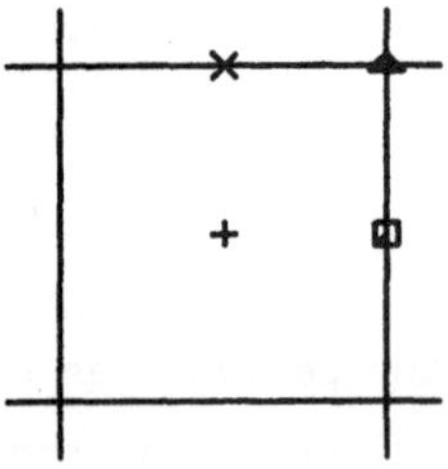

	Location	Dependent-variables	Equation
	◻	u, p_x	(2.a)
	+	v, p_y	(2.b)
	×	p	(2.c)
	▲		(2.d)

Figure 1: A computational cell, the location of the dependent variables and the points where the governing equations are satisfied.

It should be noted that the extended system of equations (1.a) - (1.d) is of the same order as the original system (1.a) - (1.c) and therefore no additional boundary conditions are needed. Equation (2.d) is satisfied only at inner node points and therefore no boundary values on the pressure gradients are needed during the iterations (see below).

The system of algebraic equations (2) is solved by a MG method similar to that which is described in reference 4. As restriction operator we use area averaging and for prolongation, linear interpolations for the corrections are used. The only differences between the PGA and the DGS schemes are in the relaxation (smoothing) step. These relaxation procedures for both methods are described in the following.

THE SMOOTHING PROCEDURE

Efficient smoothing (relaxation) of the high frequency error components is fundamental and crucial for MG methods. High frequency components are those which can be resolved on a given grid but are not resolved on any of the coarser grids. It is also important that the smoothing operator does not amplify low frequency error components. In some cases such a small amplification is acceptable provided that the amplification is compensated for on the coarse grids and totally during each MG cycle all the error-components are attenuated.

Each momentum equation can be relaxed for the respective velocity component. Multi-Grid relaxation of equations of diffusion-convection types have been considered previously [5,6]. In many cases the simplest and the most cost-effective relaxation procedure is that of Successive Point Relaxation (SPR) type. For large Re the direction of the relaxations should be aligned as much as possible, with the flow direction. For non-uniform meshes and non-simple flows, efficient smoothing may be obtained by other schemes (such as 'Convective Line Relaxations' or some other scheme which is described in [5] and [6]).

In the current MG solver only the SPR scheme was used, with a marching direction aligned with the main portion of the flow-field. The efficiency of this operator was tested for the cavity and the separator problems. It was found that the convergence rate of the MG program for the convection-diffusion equation for these cases was acceptable.

The continuity equation does not have one, natural dependent variable which can be updated during the relaxation process. An approach which was developed by Harlow and Welch [7], was to replace the continuity equation by a Poisson equation for the pressure. This equation can be derived by taking the gradient of the momentum equations and using the continuity equation. The new system of equations is of order 6 (in two-space dimensions) compared to order 4 of the original system (1). Thus, additional conditions on the boundaries must be provided. In the following we adopt the relaxation approach of Brandt and Dinar [1], to relax the continuity equation directly.

The Distributive Gauss-Seidel scheme

When the momentum equations are relaxed it is assumed that the flow field may be corrected, locally, by an irrotational field. In such cases a correction potential χ, may be introduced:

$$\Delta u = \partial_x \chi$$

$$\Delta v = \partial_y \chi \ . \tag{3}$$

With assumption (3) the continuity equation becomes:

$$(\partial_x^2 + \partial_y^2)\chi = - \partial_x u - \partial_y v \ . \tag{4}$$

In the DGS scheme a particular choice of χ is used, i.e. $\chi = a(\neq 0)$ only at the given node point which is updated in the SPR sweep, and $\chi = 0$ otherwise. The corrections to the dependent variables at each point during the point relaxation of the continuity equation, are given by:

$$\Delta u = \partial_x \chi \tag{5.a}$$

$$\Delta v = \partial_y \chi \tag{5.b}$$

$$\Delta p = Q\chi \tag{5.c}$$

When relations (5) are satisfied everywhere, the change in the residuals
of the momentum equations are given by

$$(Q \text{ grad} - \text{grad } Q)_x .$$

These changes in the residuals are small when Re is small enough or in general when the grad and the Q operators commute.

The DGS method was found to be efficient for many test problems, but for
more real flow cases, difficulties occured for non-small Re. It was found
that the reason for this depends on the underlying assumptions of the
method: the linearization of the convective terms in the Q operator and
the commutativity assumption. The last assumption was found to be the
most restrictive and therefore the PGA scheme was developed to avoid this
assumption.

The Pressure Gradient Averaging scheme

As already mentioned, this scheme uses the pressure gradients as dependent
variables. In the relaxation procedure the same steps are done as in the
DGS scheme except that when the continuity equation is relaxed, the velocity vector is updated according to relation (3). The residuals of the
momentum equations are changed and high frequency error components are
introduced. These error components are smoothed out by updating the pressure gradients. To satisfy equation (2.d) a (local) ´stream-function´ ψ is
introduced:

$$\Delta(p_x) = \partial_y \psi$$

$$\Delta(p_y) = - \partial_x \psi . \tag{6}$$

By this definition the correction problem (2.d) at each cell corner
becomes

$$(\partial_x^2 + \partial_y^2)\psi = \partial_y(p_x) - \partial_x(p_y) . \tag{7}$$

At each (inner) node point a correction ψ is computed. The pressure gradients at neighbouring point are updated according to (6), namely;

$$\Delta(p_x)|_{i,j} = -\delta/h_y \; ; \; \Delta(p_x)_{i,j+1} = \delta/h_y$$

$$\Delta(p_y)|_{i,j} = -\delta/h_x \; ; \; \Delta(p_y)_{i+1,j} = \delta/h_x$$

$$(8)$$

where δ satisfies equation (7) at the node point i,j.

The other components of the PGA-MG solver are identical to that of the DGS-MG method, with the addition that the coarse grid-corrections of the pressure gradients are interpolated linearly to the fine grid.

The PGA-MG scheme as it is described above needs more storage compared to the DGS-MG method. This additional storage can be avoided if the pressure gradients are used only during the iterations, while the pressure itself is being stored. The pressure can be updated during the relaxation sweep on equation (7): When equation (7) is satisfied at some node point, the pressure which is defined at the center of the computational cells, is updated in the four cells that share the given node points. The updating is done by integrating the corrections to the pressure gradients. It is noted that these corrections to the pressure are unique since equation (7) is satisfied for these cells. In our computational experiments which are presented in the following, the pressure gradients have been used and stored explicitly.

NUMERICAL EXPERIMENTS

Both the DGS-MG and the PGA-MG methods were applied to three test problems for different Re. The most 'smooth' test problem (Problem I) was a non-physical case with the following exact solution:

$$U(x,y) = \sin(y-x)$$

$$V(x,y) = U(x,y)$$

$$P(x,y) = \sin(y-1/2) * \sin(x) \; .$$

The steady driven cavity problem was used as Problem II. A typical high Reynolds number (Re=10^4) streamline pattern is shown in Figure 2. A less 'smooth' flow field is found in a simplified (planar) separator geometry (Problem III) where fluid is injected through one hole and is allowed to escape through some another hole (see Figure 3).

H= 0.0156
Re= 10000.0

H= 0.0156
Re= 100.0

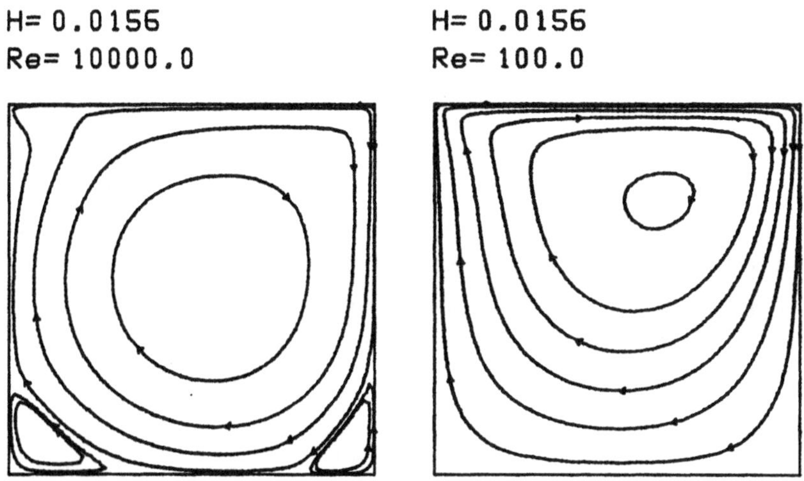

Figure 2: The streamlines in the driven cavity for Re=100 and Re=10^4.

H= 0.0313
Re= 250.0

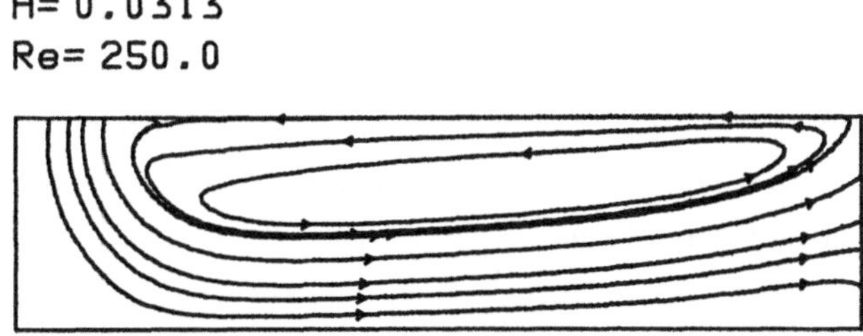

Figure 3: The geometry and the streamline in the separator, Re=250.

Relatively large separation bubles and velocity gradients are obtained with increasing Re. The boundary conditions for this problem are as follows:
i.   No-slip conditions on solid boundaries.
ii.   A given velocity profile at the entrance hole, and

iii. A fully developed velocity profile at the outflow hole.

Work-units are used as a measure for computational effort. A work-unit is defined to be equivalent to one relaxation sweep on all the equations on the finest grid. By this definition, different work-units are associated with each scheme. In the current computer code a PGA work-unit equals about 1.3 DGS work-units. The averaged convergence factor which is used to characterize computational efficiency is defined as the mean reduction of a mean residual of the system (2) by a work-unit.

The convergence histories of the numerical methods in the MG node are considered. Figure 4 shows the dependence of the convergence on the number of levels, with a given coarsest grid (and $Re=10^4$) for the DGS-MG scheme. The corresponding results for the PGA-MG scheme are shown in Figure 5. It is noted that for both methods the number of levels and the size of the mesh in the finest grid, do not effect the convergence. The averaged convergence factor is better for the PGA-MG scheme. In terms of CPU-time, however, the PGA scheme is not much more efficient than the DGS scheme.

The convergence of the methods for different Re is considered next. For Problem I, the DGS-MG scheme has a fairly constant convergence factor for $Re<10^4$ (see Figure 4). For $Re=10^5$ the convergence is slower and for $Re=10^6$, divergence occured. Figure 6 displays the convergence of the PGA scheme for the same problem. Converged solutions, with a good averaged convergence factor could be computed even for $Re=10^6$.

More realistic computations show that the results which are obtained for simple test problems are not always representative. When the DGS-MG was applied to the driven cavity (Problem II), it was observed that fast convergence was obtained for low and intermediate Re (Figure 7) but no solution could be computed for Re larger than several thousand. Even for $Re=1000$ only the initial convergence was fast, with very slow asymptotic convergence. With the PGA-MG scheme solutions up to $Re=50000$ were computed. The convergence histories for $Re < 5000$ are shown in Figure 8.

The flow field in Problem III depends very much on Re. For small Reynolds numbers only a small separation buble is observed, while for large Re (the case for $Re=250$ is shown in Figure 3), a large separated region is seen.

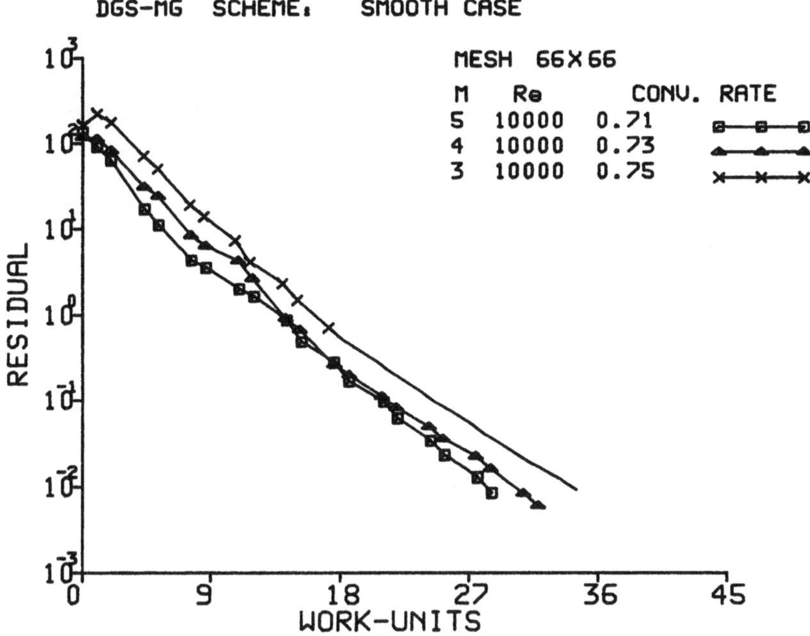

Figure 4: Convergence histories of the DGS-MG scheme for Problem I, Re = 10^4 on different grids.

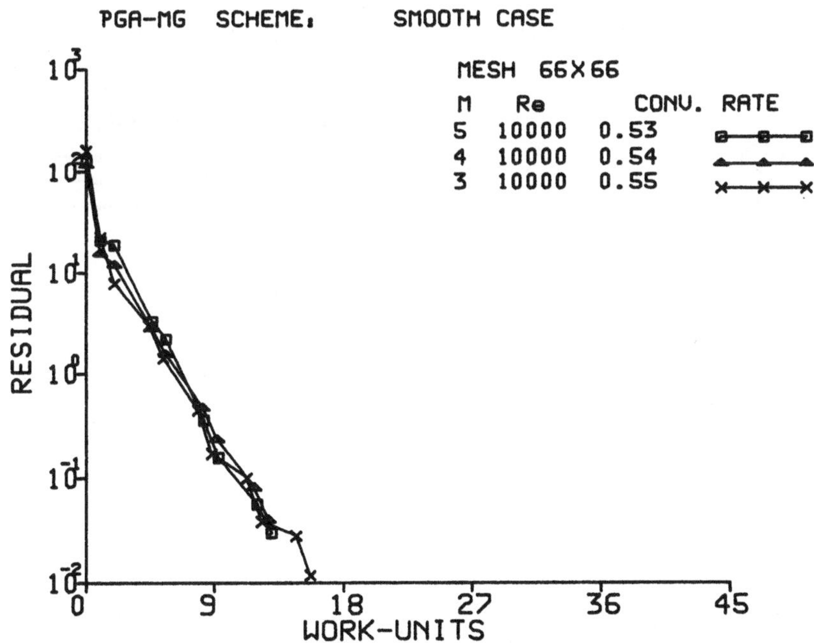

Figure 5: Convergence histories of the PGA-MG scheme for Problem I, Re = 10^4 on different grids.

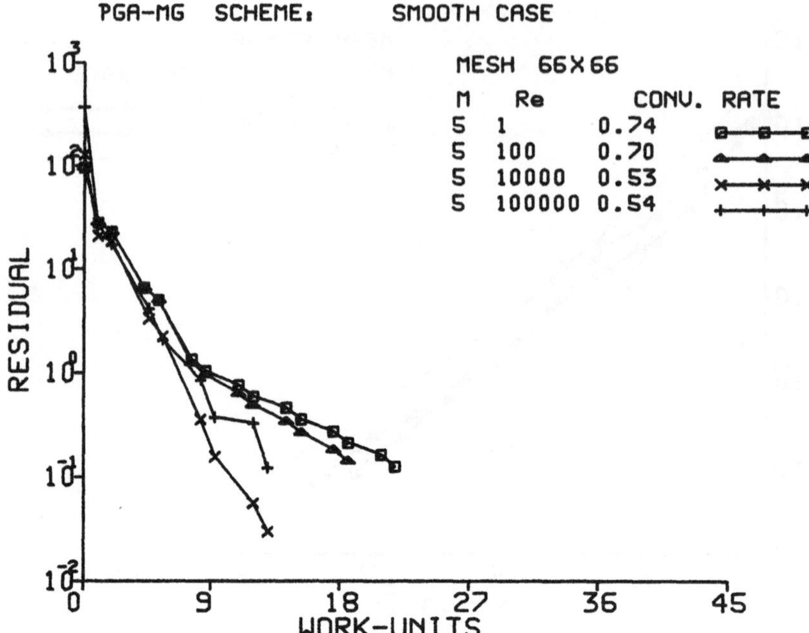

Figure 6: Convergence histories of the PGA-MG scheme for Problem I and some Reynolds numbers.

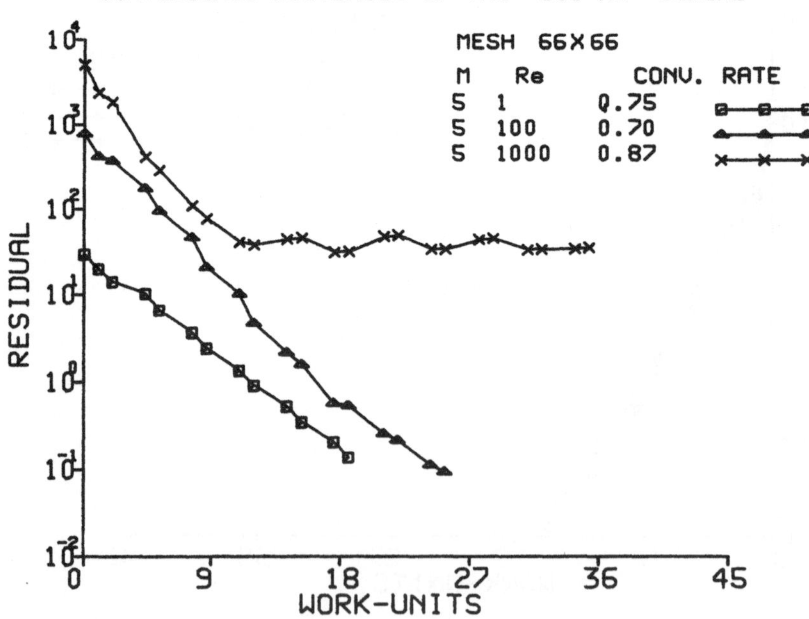

Figure 7: Convergence histories of the DGS-MG scheme for Problem II and some Reynolds numbers.

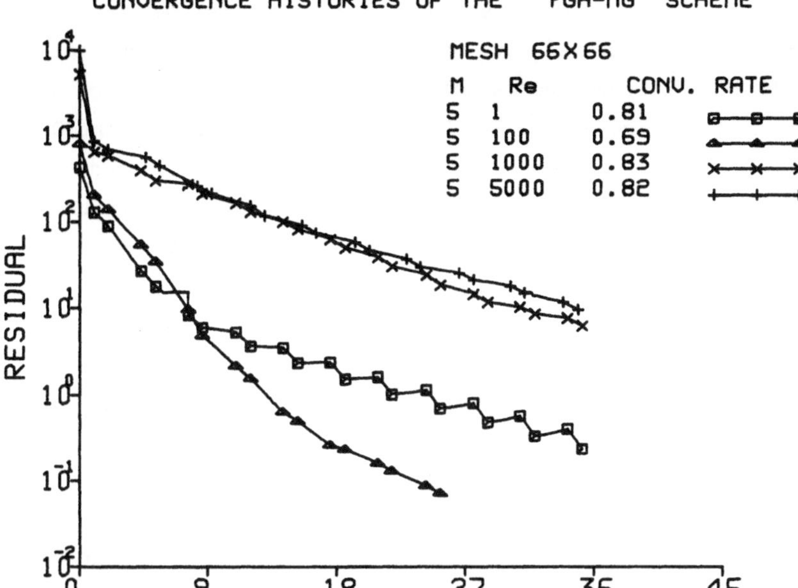

Figure 8: Convergence histories of the PGA-MG scheme for Problem II and some Reynolds numbers.

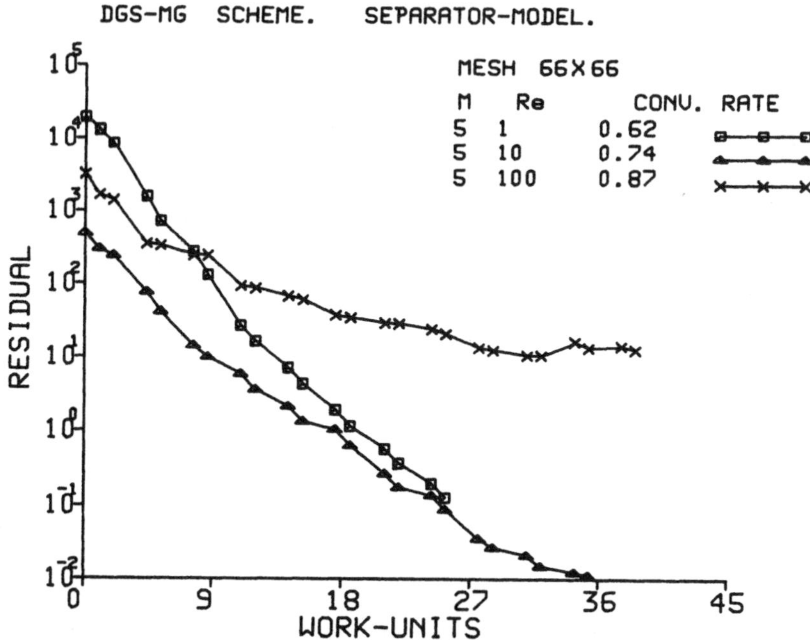

Figure 9: Convergence histories of the DGS-MG scheme for Problem III and some Reynolds numbers.

Beside the main vortex, there are two smaller vortices. Due to the extent of the main vortex large velocity gradients occur in the flow field. As is expected the DGS-MG scheme works well for low Re, but it is much slower for larger Re (see Figure 9). For Re larger than about 150, the DGS method diverged. The PGA-MG method, on the other hand converged for Re up to several hundreds, with the separated region extending up to the outflow boundary. The convergence histories of the PGA scheme for some Reynolds numbers are shown in Figure 10.

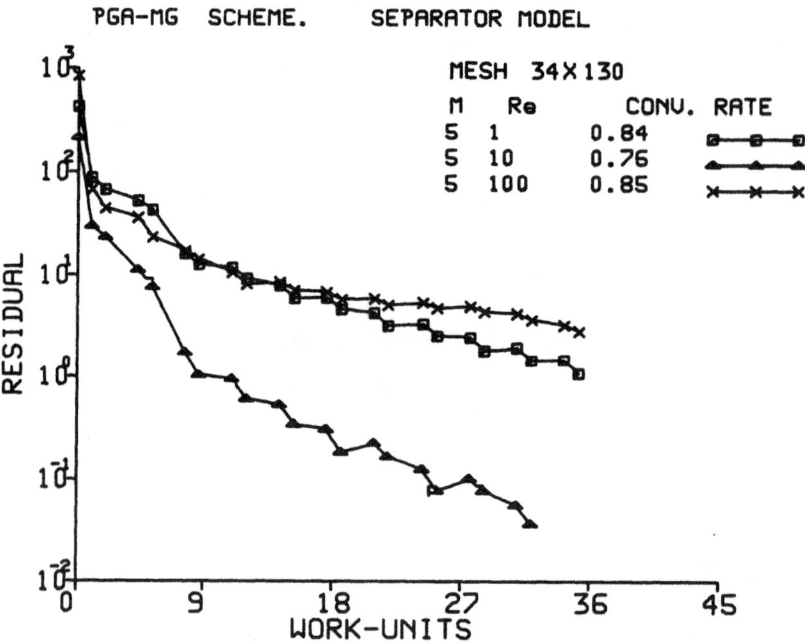

Figure 10: Convergence histories of the PGA-MG scheme for Problem III.

CONCLUDING REMARKS

A new scheme for the incompressible Navier-Stokes equations was presented. The new method was tested, together with the DGS scheme on three different problems. Both methods showed fast convergence for small Reynolds numbers. For larger Re the convergence factor of the PGA scheme was better than that of the DGS scheme. With increasing Re, solution could be computed only by the PGA scheme. The new PGA scheme, in the present form needs larger computer memory, but as it was indicated this difficulty can be easily overcome.   From the numerical experiments we concluded that for problems where the DGS scheme could maintain a convergence factor of about 0.6, it was competative with the PGA scheme. On the other hand, for large Re the later scheme had to be used, although that scheme itself was slower than for smooth cases.

Our extensive computational experiments showed that numerical schemes have to be tested for more realistic problems rather than only on smooth test cases. Usually, most problems occur for flows with large gradient or for larger Re when some of the linearizations are not accurate. Since the separator problem has a simple geometry, it is believed that it can be used as a standard problem for testing the robustness of new numerical schemes.

REFERENCES

1.   Brandt, A. and Dinar, N.: Multigrid solution to elliptic flow problems, in Numerical methods in PDE, S.V. Parter ed. pp. 53-147, Academic Press (1979).

2.   Thunell, T and Fuchs, L.: Numerical solution of the Navier-Stokes equations by Multi-Grid techniques, in Numerical methods in laminar and turbulent flow, C. Taylor and A.B. Schrefler eds. pp. 141-152, Pineridge Press (1981).

3.   Fuchs, L. and Zhao, H-S.: Solution of three-dimensional viscous incompressible flows by a multi-grid method Numerical Methods in Fluids (1984) to appear.

4.   Fuchs, L.: New relaxation methods for incompressible flow problems, in Numerical methods in laminar and turbulent flow, C. Taylor, J.A. Johnson and W.R. Smith eds. pp. 606-616, Pineridge Press (1983).

5.   Fuchs, L.: A fast numerical method for the solution of boundary value problems, TRITA-GAD-4, ISSN 0281-7721 (1980).

6.   Fuchs, L.: Multi-Grid solution of the Navier-Stokes equations on non-uniform grids, in NASA CP-2202, pp. 83-100 (1981).

7.   Harlow, F.H. and Welch, J.E.: Numerical calculation of time-dependent incompressible flow, Physics of Fluids, Vol. 8, pp. 2182-2187 (1965).

DEFECT CORRECTIONS AND HIGHER NUMERICAL ACCURACY

L. Fuchs
Department of Gasdynamics
The Royal Institute of Technology
S-100 44 Stockholm, Sweden

SUMMARY

Different variants of defect-correction techniques were incorporated into a basic Multi-grid solver of the vorticity transport equation. A scheme, which uses a sequence of finite-difference approximations and only one defect-correction step with each approximation, has been tested. The asymptotic order of convergence of the solution with this method can be estimated. The main advantage of this scheme is its improved stability properties compared to the classical method of defect-correction iterations. The levels of the errors for some finite-difference approximations and for some accuracy improving schemes, are compared. The numerical experiments show that the accuracy of well proven and robust numerical codes can be improved by simple means and with relatively small increase in computational effort.

INTRODUCTION

Numerical methods for the solution of partial differential equations can be characterized by the efficiency of the algorithm and the accuracy of the numerical results. Usually, these two elements are connected and the accuracy of the results is determined by the amount of available computer time (and memory). For this reason it is of great importance to have both an efficient (algebraic) solver to the finite difference equations (FDE) and schemes which offer higher accuracy without a significant increase in the computer time.

In this work we consider the accuracy levels of the numerical approximations to the solution of the vorticity transport equation. Such accuracy levels must always be computed whenever the numerical results are to be used for quantitative comparison with experiments. Three test problems simulating three different degrees of flow smoothness, are solved. The same basic Multi-Grid(MG) solver is used in all the cases. This solver has formally only first order accuracy since it uses up-winded differences to approximate the convective terms. The up-winded differencing introduce artificial viscosity which increases the stability and decreases the accuracy of the scheme. The impact of using low order schemes can be fully appreciated only if the level of accuracy can be estimated. When the computed level of accuracy is not good enough then either finer meshes or higher order schemes, or both must be used. Using a finite-difference scheme of higher order of accuracy, on a given mesh, does not neccessarily increase the accuracy level of the solution (see below). Secondly, higher order schemes are usually, less stable in an iterative process compared to lower order finite differences.

Recently [1], the numerical simulation of the flow in a driven polar cavity was compared with experimental results. It was found that for low Reynolds numbers (Re≈60), solutions on coarse grids (of about 20x20 uniformly distributed intervals) had a level of accuracy comparable with the accuracy of the experimental results (2-3%). For high Re (about 350) numerical solution on fine grids (with more than 10^4 node points) was not accurate enough (the mean error was about 10%). With this observation in mind it is clear that higher order methods,which would provide higher levels of accuracy, are desirable. Furthermore, we would like to implement these FDE in a simple and stable manner. Several such possibilities are discussed in the following.

The method of Defect Correction (DC) has been used to improve the order of accuracy of numerical schemes. Defect corrections in the MG context, including some convergence proves, have been discussed by Hackbusch [2] (and the references there). Brandt [3] suggested the use of the 'double discretization'(DD) scheme as an integral part of the MG scheme. These methods, and a new variant of defect correction (the Multi-Stage one-step Defect Correction, MS-DC) scheme, are described in the following. The later method is fast and offers increased stability compared to the standard DC techniques. The number of DC steps to achieve the level of accuracy comparable to that of the exact solution to the corresponding FDE, is estimated.

The different accuracy-improving schemes have been implemented and used to

solve three test problems for different values of Re. The accuracy levels
of the different solutions are compared and the 'asymptotic' convergence
rates are computed and compared with the expected ones.

THE MODEL PROBLEM

The level of the error of a given numerical method can be estimated only if
solutions are computed on a sequence of successively refined grids. From
these solution an extrapolated solution to 'zero' mesh, may be found and
'the errors' may be estimated. However, such extrapolations are not always
valid and therefore, for simplicity we use here the following explicit
forms for the streamfunction:

a. 'Smooth case':

$$\psi_1 = \sin(cx + y)$$

'Cavity like flow':

$$\psi = \{1 + XRE[(x - x_0)^2 + (y - y_0)^2]\}\alpha + \beta \, \sin \pi x \, \sin \pi y$$

where $\alpha = x^2(1-x)^2 \, y^2(1-y)$

b. With $\psi_2 = \psi$ and XRE = 1 simulating low Reynolds number flows
   and

c. $\psi_3 = \psi$ and XRE = 50 simulating high Reynolds number cases.

These functions are exact solutions to the vorticity transport equation

$$L\omega = R \tag{1}$$

where

$$L = Re[u\partial_x + v\partial_y] - \nabla^2$$

and

$$\omega = u_y - v_x \; ; \; u = \psi_y \, , \, v = - \psi_x$$

where R is choosen in such a way that ω satisfies equation (1).

Equation (1) is discretized on a uniform mesh with mesh spacings h in both space directions. The differential operator L in (1) is approximated by either L_1, L_2 or L_4, where

$$L_1 = \mathrm{Re}[u\overset{\leftarrow}{\partial}_x + v\overset{\leftarrow}{\partial}_y] - \nabla^2_5$$

$$L_2 = \mathrm{Re}[u\tilde{\partial}_x + v\tilde{\partial}_y] - \nabla^2_5$$

$$L_4 = \mathrm{Re}[u\hat{\partial}_x + v\hat{\partial}_y] - \nabla^2_9$$

where $\quad \nabla^2_5 \Phi = \dfrac{1}{h^2}(\Phi_{i+1\,j} + \Phi_{i-1\,j} + \Phi_{ij-1} - 4\Phi_{ij})$

$$\nabla^2_9 \Phi = \frac{1}{12h^2}[-\Phi_{i-2j} - \Phi_{i+2j} - \Phi_{ij-2} - \Phi_{ij+2} + 16(\Phi_{i-1j} + \Phi_{i+1j} + \Phi_{ij-1} + \Phi_{ij+1}) - 60\Phi_{ij}]$$

$$\alpha\overset{\leftarrow}{\partial}_{\cdot}\Phi = \alpha\frac{v}{h}[\Phi_{\cdot} - \Phi_{\cdot-v}] \qquad v = \mathrm{sign}(\alpha)$$

$$\alpha\tilde{\partial}_{\cdot}\Phi = \frac{\alpha}{2h}[\Phi_{\cdot+1} - \Phi_{\cdot-1}]$$

$$\alpha\hat{\partial}\Phi = \frac{\alpha}{12h}[8(\Phi_{\cdot+1} - \Phi_{\cdot-1}) - (\Phi_{\cdot+2} - \Phi_{\cdot-2})] .$$

L_1, L_2 and L_4 have a formal accuracy of first, second and fourth order, respectively.

NUMERICAL SOLUTION PROCEDURES

Denote the basic finite-difference equations corresponding to equation (1) by

$$L_h \bar{\omega} = R. \tag{2}$$

This problem is solved by a MG-solver which uses a line relaxation scheme, weighted residual transfer and linear interpolations for the corrections. The basic MG solver had a satisfactory convergence rate for all the cases which were tested. The same MG solver, with some minor modifications corresponding to the different methods, was used to compute all solutions.

DEFECT CORRECTIONS METHODS

The classical method of defect corrections has been used both to improve accuracy (see e.g. Hackbuschs [2]) and for the solution of 'perturbed' problems (see e.g. Martin [4]) or non-linear problems (Fuchs [5]). The basic iterative DC process to improve accuracy is defined by

$$L_1 \omega^{(1)} = R$$

$$L_1 \omega^{(i)} = L_1 \omega^{(i-1)} - L_k \omega^{(i-1)} + R \quad ; \quad i > 1 \tag{3}$$

where L_1 and L_k are approximations to L. When the iterative process converges the solution $\omega^{(i)}$ is an approximation to $\tilde{\omega}$ which solves the equation $L_k \tilde{\omega} = R$.

The method of DC is very simple to implement, but the algebraic convergence i.e. the rate at which $\omega^{(i)}$ converge to $\tilde{\omega}$, may be very slow. Furthermore, if L_k is a high order approximation to L, the iterative process may have a tendency to diverge due to low-frequency instability.

These difficulties can be avoided if one utilizes the following factors:
i.   The level of algebraic convergence need not be much better than the accuracy limit due to the discretization, and
ii. Low order discrete schemes are more stable, and therefore they should be used whenever they do not impair the accuracy.

With these observations in mind we construct the following variant of defect correction procedure: Consider a sequence of N approximations $\{L_j\}$, where each element in the set approximates L with an accuracy of order p_j (e.g. truncation error of $O(h^{p_j})$). Assume that the sequence is ordered such that $p_j \leqslant p_{j+1}$; $1 \leqslant j < N$. A multi-stage one-step defect-correction (MS-DC) scheme is defined as

$$L_1 \omega^{(1)} = R$$

and

$$L_1 \omega^{(j)} = R + L_1 \omega^{(j-1)} - L_j \omega^{(j-1)} \qquad 1 < j \leqslant N \ . \tag{4}$$

Consider the formal accuracy of the solution $\omega^{(j)}$, and define the error,

$e^{(j)}$, as

$$e^{(j)} = \omega - \omega^{(j)} \tag{5}$$

where $L\omega = R$.

If the error can be represented by the lowest order term in h, denote this order by q_j (i.e. $e^{(j)} = O(h^{q_j})$).

Further, if L_j can be linearized without introducing lower order errors and if each DC step is solved accurately enough, then

$$L_j e^{(j)} = L_1 e^{(j-1)} - L_j e^{(j-1)} + O(h^{p_j}) \tag{6}$$

which gives that

$$q_j = \min\{p_1 + q_{j-1}, \ p_j + q_{j-1}, \ p_j\} \tag{7}$$

The corresponding relations for the classical DC scheme (3) is given by

$$\bar{q}_i = \min \{p_1 + \bar{q}_{i-1}, \ p_k + \bar{q}_{i-1}, \ p_k\} \ . \tag{8}$$

By relation (7) we observe that the order of the approximation, p_j, in the j-th step should be

$$p_j \geq p_1 + q_{j-1} \ . \tag{9}$$

When the sequence of approximations $\{L_j\}$ is ordered so that relation (9) is satisfied, the MS-DC scheme is stable if all the approximations do not have common unstable modes.

The number of DC steps, K, that are needed to reach an accuracy of $O(h^{p_N})$, with the target operator L_N, can be estimated by

$$K = p_N/p_1 - 1 \ . \tag{10}$$

This number of DC steps is neccessary both for schemes (3) and (4), and is adequate only if the assumptions which are stated above are valid. When the truncation errors cannot be represented by the lowest order term, the convergence of the DC schemes is slower. In such cases one may use multiple

DC steps with the low order approximation, until a satisfactory level of accuracy is obtained, and than continue with the MS-DC steps. The use of standard DC scheme in such cases, may result in non-convergence or even divergence.

THE DOUBLE DISCRETIZATION SCHEME

A DC scheme in the Multi-grid context was suggested by Brandt [3]. The DC step is applied only during the transfer of the residuals to coarse grids, k, on which a problem of the following form is treated:

$$L_1^k \; \bar{\omega}_k = R_1^k \; .\tag{11}$$

The right hand side of the coarse grid problem (11) is computed by:

$$R_N^k = I_k^{k+1} (R_N^{k+1} - L_N^{k+1} \; \bar{\omega}_{k+1}) + L_N^k \; I_k^{k+1} \; \bar{\omega}_{k+1}\tag{12.a}$$

$$R_1^k = I_k^{k+1} (R_N^{k+1} - L_N^{k+1} \; \bar{\omega}_{k+1}) + L_1^k \; I_k^{k+1} \; \bar{\omega}_{k+1}\tag{12.b}$$

where I_k^{k+1} denotes the transfer of the residuals from the fine grid k+1, to the coarse grid, k. The DC step on a given grid is defined by relation (12.b) whereas (12.a) defines the residual of the target operator and which is transferred to even coarser grids.

The accuracy which is obtained by this ('double discretization'-DD) scheme depends not only on the operators L_1 and L_N but also on the basic elements of the MG scheme; that is, the residual transfer, and the related restriction and prolongation operators. The smoothing properties of the relaxation process of equation (11) should be such that the modes related to the lower accuracy of the relaxation scheme shall not be resolved during the smoothing. This implies that e.g. fast solvers cannot be applied on (11) on all grids. The analysis of double discretization schemes requires therefore a two-level smoothing analysis similar to the one used by Hackbusch [2].

It should be emphasized that the solution obtained by the DD method is not an algebraic solution to the target FDE. The solution is an approximation to the differential solution with an accuracy $O(h^q)$, such that

$p_1 < q \le p_N$. Furthermore, the approximation is non-smooth and therefore applying on it a single DC-step does not result in an improved accuracy level.

In our numerical experiments, we applied the MS-DC and the DD schemes on the three test problems for some Reynolds numbers. A comparison of the numerical accuracy of the results is given in the following.

NUMERICAL EXPERIMENTS

In our numerical experiments we compared the accuracy levels of the different FDE and the accuracy of the methods described above. Here, we define the accuracy of the FDE, L_k, by solving

$$L_1\hat{\omega} = R + L_1\omega - L_k\omega \ . \tag{13}$$

When the exact solution, ω, is not known, one has to use a DC scheme to find $\tilde{\omega}$, which solves $L_k\tilde{\omega} = R$. The difference between $\hat{\omega}$ and $\tilde{\omega}$ is $O(h^p k)$. For non-smooth problems the level of this difference may be non-negligible. In such cases the accuracy of $\hat{\omega}$ is better than the accuracy of $\tilde{\omega}$. However, we use here $\hat{\omega}$ as reference for the 'best possible' solution which can be obtained with L_k and a given problem.

First we consider the smooth case ($\psi = \psi_1$). A comparison of the accuracy levels of the solution with L_1, L_2 and L_4 (in the sense of solving equation (13)) is shown in Figures 1-3 for Re = 10, 100 and 10^5, respectively. The computed ('asymptotical') order of convergence is found by using the four finest grids and using a (RMS-best) line fitting. It can be observed that the accuracy level of the solution to the fourth order operator is better than that of the second order, only for fine enough meshes. The 'breaking-even-point' (i.e. the mesh for which the level of the accuracy for both FDE is the same) moves from a mesh with 16 intervals to a mesh with 64 intervals as Re increases from 10 to 10^5. The computed order of convergence reaches the theoretical convergence order on fine meshes and for low and moderate Re.

The sequence of solutions which was computed by the MS-DC scheme (4) results in accuracy levels which are shown in Figure 4. The computed order of convergence is as expected by relation (7). Furthermore, the levels of accuracy are almost the same as those obtained by solving equation (13) (see Fig. 3). Similar results were obtained for moderate Re, but for larger values of Re the computed convergence does not reach the theoretical one, even for the finest meshes. A similar behavior is noted for the other test cases. The MS-DC results for the 'low-Reynolds-number' cavity-like case are shown in Figures 5 and 6. For Re = 100 the computed convergence order and the levels of the error are very close to that of the 'exact' solution, $\hat{\omega}$. For bigger Re (as in figure 6, Re = 10^4) the convergence order is still according to relation (7), but the error levels of the MS-DC are larger than the corresponding 'exact' solutions to the FDE. A similar result is obtained with 'high-Reynolds-number' cavity-like test case. Figure 7 shows the MS-DC error levels for Re = 100, with the expected convergence orders. The levels of the errors on the finest grid, are however, larger by about an order of magnitude compared to the 'exact' solution to the FDE.

The DD scheme, with a second order and a fourth order operator was incorporated into the basic MG solver. For both cases the same schemes for the residual transfer and interpolation of the corrections, were used. Since only linear interpolations were employed, no full effect from the fourth order FDE could be expected. In the following we compare the accuracy of these DD schemes with the 'exact' and the MS-DC results. Figures 8-10 show the errors with the different methods for three cases respectively: smooth case and Re = 10, cavity-like case (XRE = 1) and Re = 100, and cavity-like case (XRE = 50) and Re = 100. These figures show that the DD-schemes result in a better level of accuracy than that of the MS-DC scheme to second order, and that fourth order DD is not much better than its second order counterpart, if the MG method is not adapted to the higher order FDE. After a DD solution, the approximation contains low-amplitude error components but with high frequencies. For this reason no significant improvement is obtained when the MS-DC scheme is applied once. A repeated use of the MS-DC scheme results in an accuracy improvement according to relation (7).

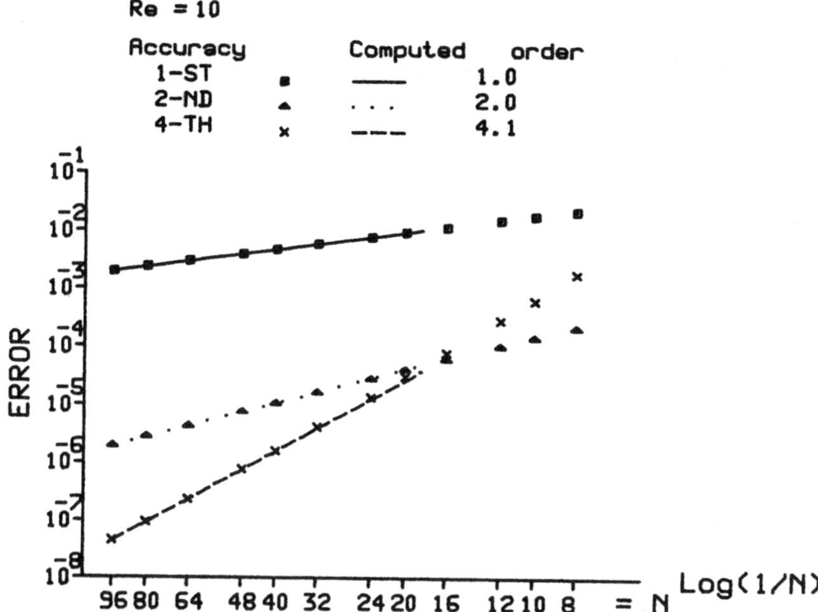

Figure 1: The error vs. the mesh spacing with three different FDE.
Smooth problem ($\psi=\psi_1$). Re 10.

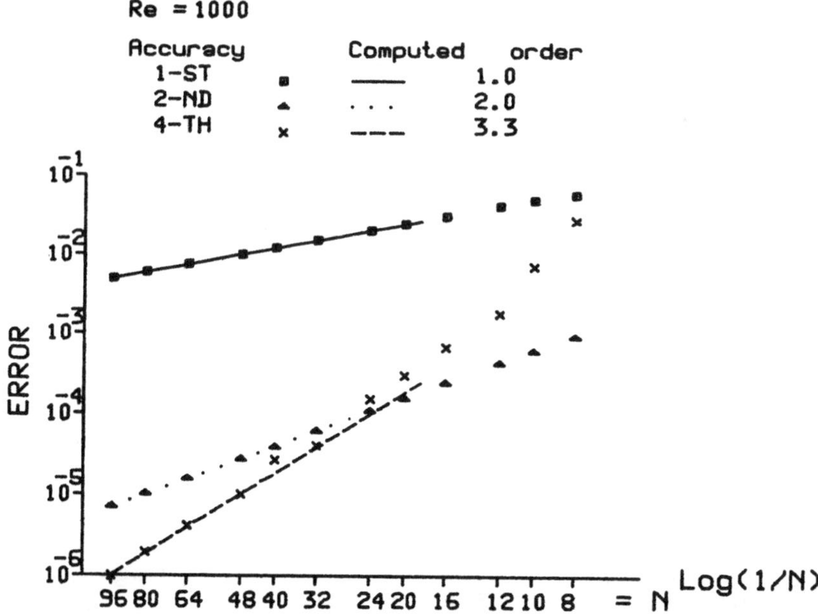

Figure 2: The error vs. the mesh spacing with three different FDE.
Smooth problem ($\psi=\psi_1$). Re = 10^3.

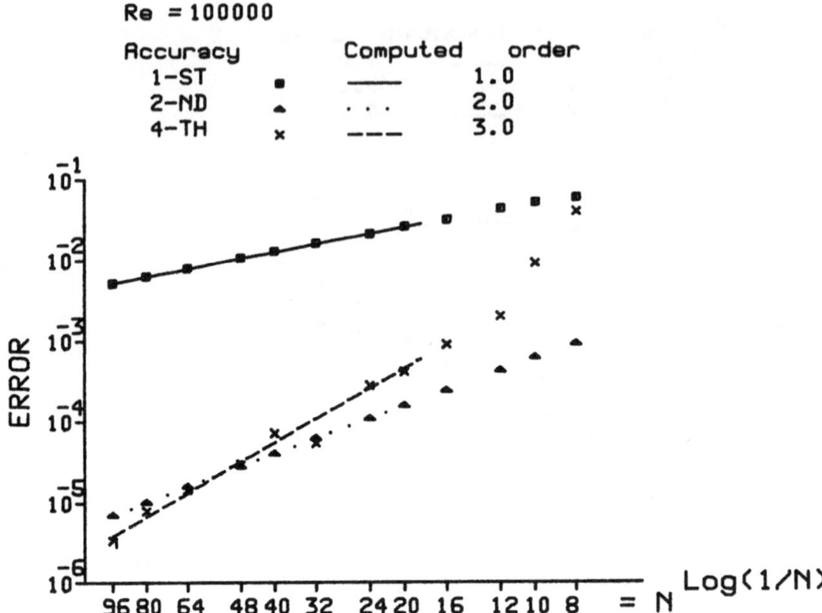

Figure 3: The error vs. the mesh spacing with three different FDE.
Smooth problem ($\psi=\psi_1$). Re = 10^5.

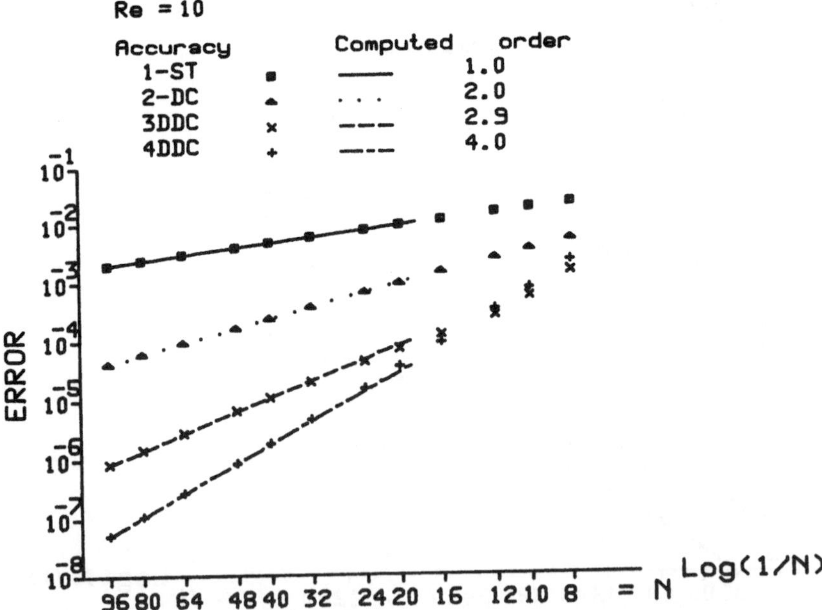

Figure 4: The error vs. the mesh spacing with the MS-DC scheme.
Smooth problem ($\psi=\psi_1$). Re = 10.

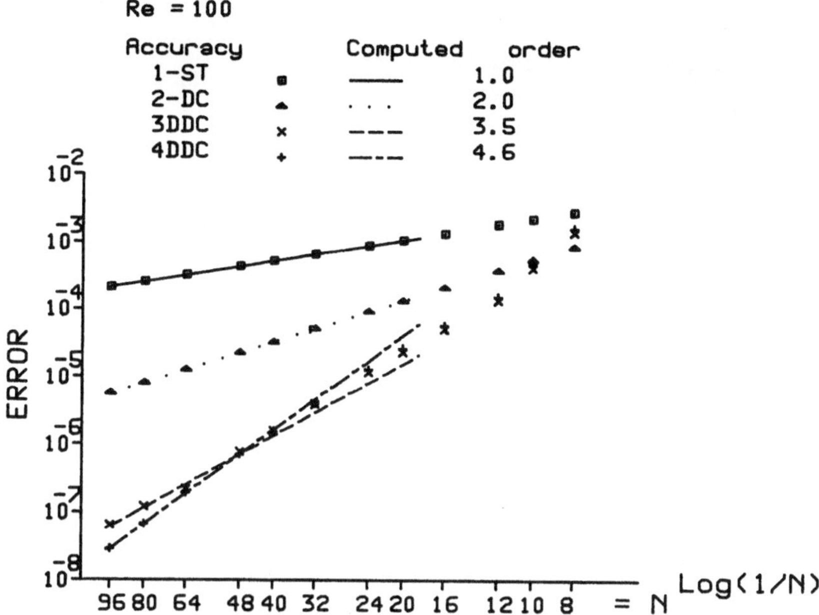

Figure 5: The error vs. the mesh spacing with the MS-DC scheme.
Cavity-like flow at low Reynolds numbers ($\psi=\psi_2$; XRE=1). Re= 100

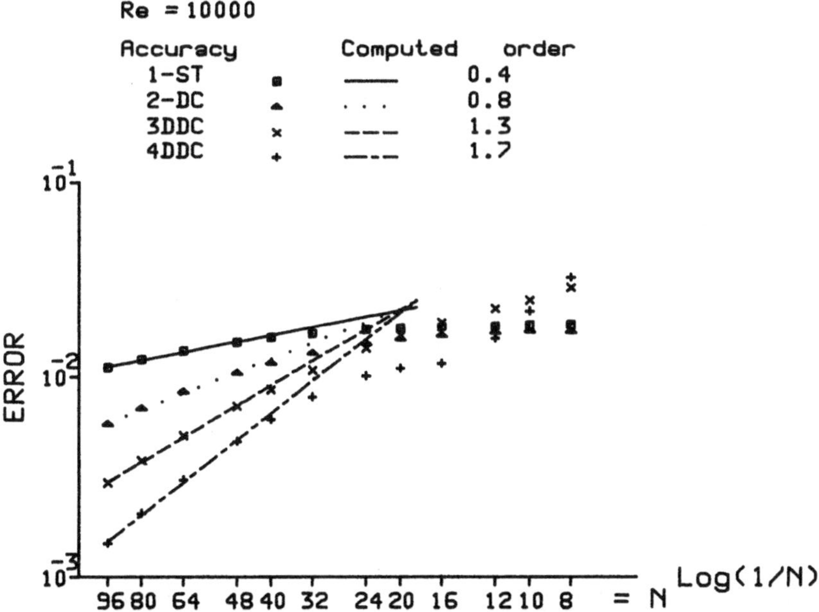

Figure 6: The error vs. the mesh spacing with the MS-DC scheme.
Cavity-like flow at low Reynolds numbers ($\psi=\psi_2$; XRE=1). Re=10^4.

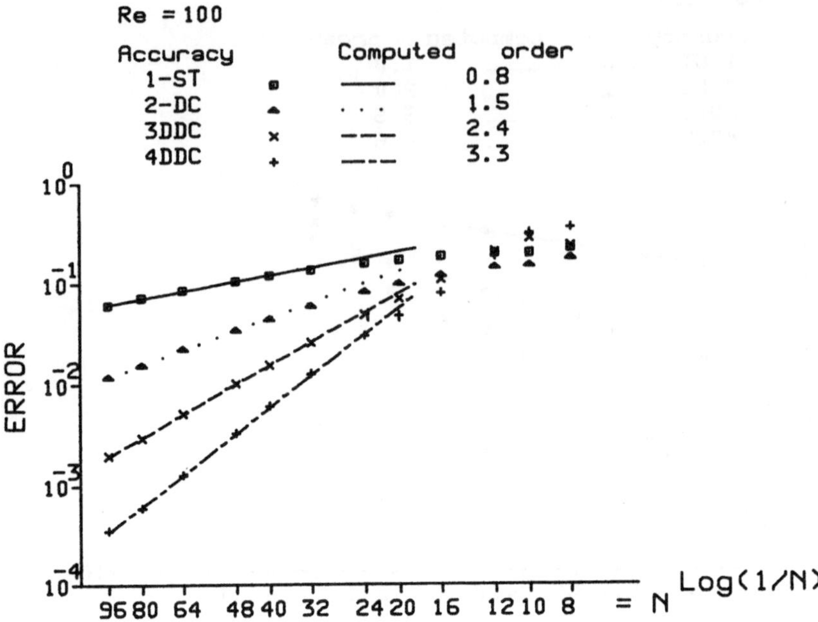

Figure 7: The error vs. the mesh spacing with the MS-DC scheme. Cavity-like flow at high Reynolds numbers ($\psi=\psi_2$; XRE=50) Re=100.

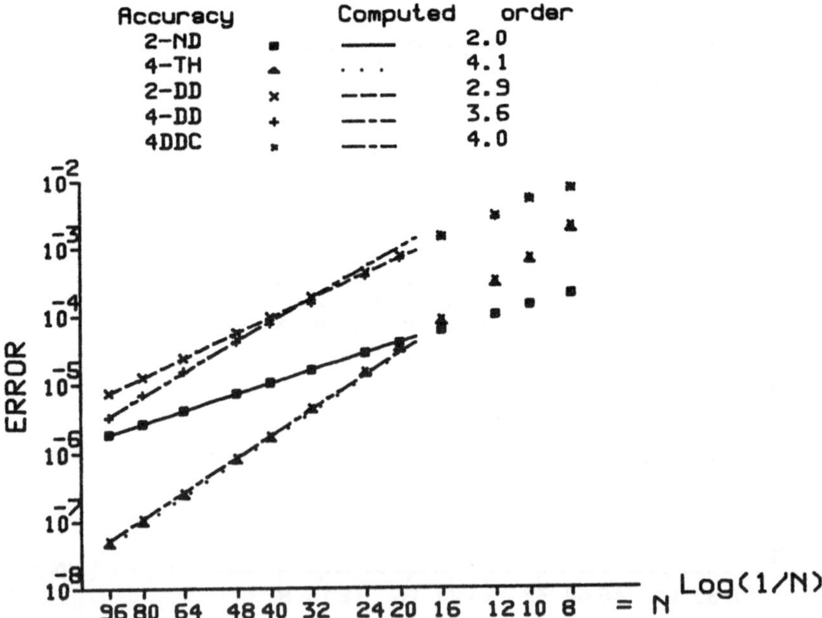

Figure 8: The error vs. the mesh spacing with two DD-schemes. Smooth problem ($\psi=\psi_1$). Re=10.

65

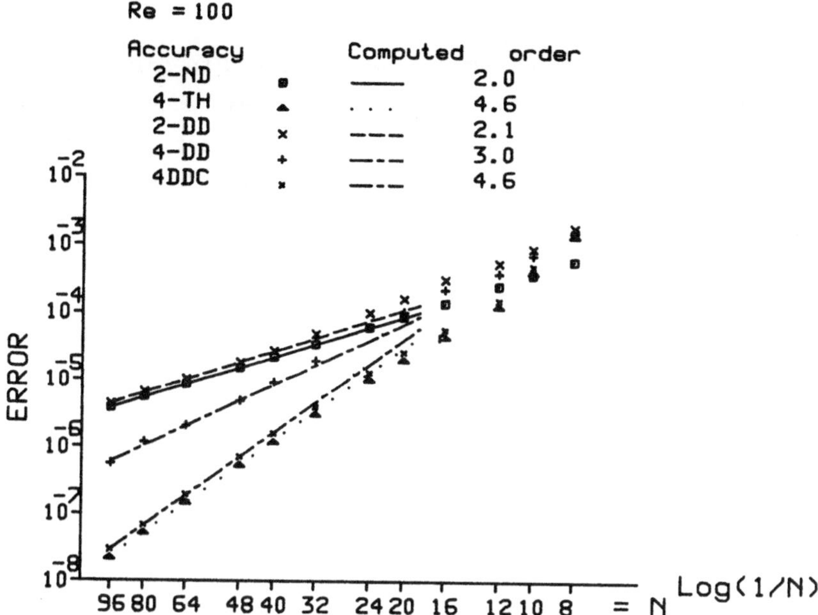

Figure 9: The error vs. the mesh spacing with two DD-schemes.
Cavity-like flow at low Reynolds numbers ($\psi=\psi_2$; XRE=1). Re=100.

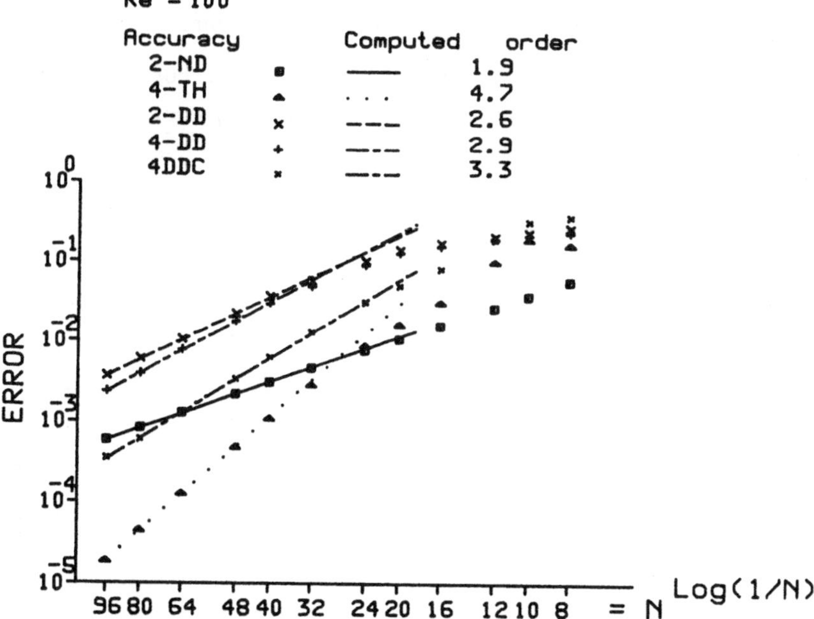

Figure 10: The error vs. the mesh spacing with two DD-schemes.
Cavity-like flow at high Reynolds numbers ($\psi=\psi_2$; XRE=50).
Re=100.

CONCLUDING REMARKS

The multi-step defect correction scheme was introduced to gain an improved
stability in the iterative DC process. The MS-DC scheme was tested on the
vorticity transport equation for some cases and different Reynolds numbers.
The accuracy of the MS-DC scheme was compared to that of the 'exact' solution to the FDE and the accuracy which may be obtained by the double-
discretization scheme. It has been found that the MS-DC scheme is very
simple to implement and the accuracy of the solutions can be improved
easily. The new scheme is more stable and requires less programming effort
compared to the DD-scheme. On the other hand it is anticipated that it may
be less efficient compared to DD-MG methods with optimal prolongation and
restriction operators. However, because of its simplicity, the MS-DC scheme
is in many cases the most feasable way of increasing the numerical accuracy
of existing numerical codes.

REFERENCES

1.	Fuchs, L. and Tillmark, N.: Numerical and experimental study of cavity
	flows. Numerical Methods in Fluids (1984), to appear.

2.	Hackbusch, W.: On multi-grid iterations with defect correction, in
	Multigrid Methods, W. Hackbusch and U. Trottenberg eds. pp.461-473,
	Springer-Verlag, Berlin (1982)

3.	Brandt, A.: Multigrid solutions to elliptic flow problems, in Numeri-
	cal methods for partial differential equations, S.V. Parter ed.
	pp. 53-147, Academic Press (1979).

4.	Martin, D.E.: Progress in application of direct elliptic solvers to
	transonic flow computations, in Aerodynamic analyses requiring ad-
	vanced computers, NASA SP-347, part II, pp. 839-870,(1975).

5.	Fuchs, L.: A Newton-multi-grid method for the solution of non-linear
	partial differential equations, in Proc. of the BAIL-I Conference,
	J.J.H. Miller ed. pp. 291-296, Boole Press, Dublin (1980).

ON A MULTIGRID METHOD TO SOLVE THE INTEGRAL EQUATIONS OF 3-D STOKES' FLOW

F.K. Hebeker[*]

Fachbereich 17 Mathematik

Universität D-479o Paderborn

SUMMARY

The 3-D exterior Stokes' boundary value problem is investigated. The classical hydrodynamical potential theory is developed up to a boundary element method, where the algebraic system is quickly solved by a two-level multigrid method. The results of numerical test calculations are discussed.

INTRODUCTION

Let us consider the 3-D exterior Stokes' boundary value problem

$$- \Delta u + \nabla p = o \ , \quad \text{div } u = o \quad \text{in } \Omega' \ ,$$

$$u_{|\partial \Omega} = g \ , \quad u_{|\infty} = o \quad \text{at infinity} \tag{1}$$

for the velocity field u and the pressure function p of a homogeneous viscous incompressible fluid flow in a domain $\Omega' = \mathbb{R}^3 \setminus \overline{\Omega}$ exterior to a smoothly bounded ($\partial \Omega \in C^{\infty}$) rigid body Ω . Recently boundary element methods have been introduced in order to manage to compute real 3-D slow viscous flows in domains of smooth but else arbitrary shape (Hsiao--Wendland-Fischer [3], Nedelec-Zhu [15], and the author [6], [7]). Numerical test calculations which reflect the whole 3-D problem have been communicated in [6],[7],[15].--For general boundary element methods we refer to [1].

---

[*]   This work has been partly supported by the Deutsche Forschungsgemeinschaft in its special program "Finite Approximationen in der Strömungsmechanik".

Parallel to these investigations the multigrid methods have been employed to the integral equations of Laplacean problems (Schippers [12], Wolff [14], Novak [1o]). Hence in the present note we try to combine both developments and intro duce a boundary element/multigrid method for the Stokes' problem (1). Our approach to solve the linear algebraic system is a two-level multigrid method, consisting of a sequence of one Gauß-Jacobi iterative step on a fine grid and one correcting step on the double-sized coarse grid. Extensions to more refined procedures as MGR methods (Ries-Trottenberg-Winter [11]) currently are under study.-- This note will be finished with the numerical results of test calculations.

The present author wishes to express sincere thanks to professors D. Braess (Bochum), W. Hackbusch (Kiel), and U. Trottenberg (Essen/Bonn) for fruitful talks and criticism.

A   BOUNDARY ELEMENT   METHOD

Stokes'differential equations form an elliptic system in the sense of Agmon-Douglis-Nirenberg, but a simple fundamental solution, a $(4,4)$-matrix, is available. We refer to [9] for the hydrodynamical potential theory. In particular the hydro-dynamical potentials of the simple layer $V\phi$ and of the double layer $W\phi$ (with vector surface source ϕ) both satis-fy the homogeneous Stokes' equations in $\mathbb{R}^3\diagdown\partial\Omega$ with cor-responding pressure potentials.

The classical way to solve (1) is by means of the poten-tial ansatz $u = W\phi$. But then we are led to integral equa-tions for which we generally have non-existence and non--uniqueness. Hence extending an idea of Leis, Brakhage, and P. Werner for Helmholtz' equation we make the mixed potential ansatz

$$u = W\phi + \eta V\phi , \quad p \text{ analogous,} \tag{2}$$

where η is a constant free parameter. The jump relations $(\rightarrow[9],[2])$ then lead to the boundary integral equations' system of the second kind for the unknown vector surface source ϕ :

$$\phi - 2W\phi - 2\eta V\phi = -2g \ , \quad \text{on } \partial\Omega \ . \qquad (3)$$

This system proves to be *uniquely solvable* for any continuous boundary data, if η is chosen negative ($\rightarrow$[6]).
When (3) is (approximately) solved, then (2) is the wanted
Stokes' flow in Ω' . The potentials $W\phi$, $V\phi$ are weakly
singular integrals on the boundary, hence (3) is an attractive
numerical approach to problem (1).

Besides, if ϕ solves (3), then the force F exerted
by the fluid flow on the rigid body quickly is computed by
$$F = - \eta \int_{\partial\Omega} \phi \ do \ , \qquad (4)$$
a surface integral only. This nice property has been pointed
out to us by T. Fischer (Darmstadt),

How to solve (3) now ? Perhaps most simple and efficient
seems to be a collocation procedure. As coordinate functions
we use globally continuous and piecewise bilinear polynomials
on the parameter space of the boundary. Then the integrals
in (3) are computed using a simple formula of numerical
quadrature. So (3) is discretized, and we are led to a linear
algebraic system called
$$(I_h - K_h)\phi_h = f_h \ , \qquad (5)$$
with a nonsparse but relatively small matrix $I_h - K_h$. We are
able to show that (ϕ_h) converges to ϕ like $h.\log\frac{1}{h}$,
and the same holds for the approximate flow fields
(u_h) asymptotically in the whole flow region ($\rightarrow$[8]).

A   TWO-LEVEL   MULTIGRID   METHOD

Usually systems like (5) are solved iteratively, and
here we propose a two-level multigrid method. To be concrete,
let us describe the procedure in the case of a spherical
rigid body - but we point out that the calculations are performed without using any symmetry properties (we only used
polar coordinates as a global coordinate frame). Hence the
method and the results are considered as typical for 3-D flows
in more general flow regions.

By means of normed polar coordinates (θ,φ) the sphere

is transformed to the unit square $E = [o,1]^2$. E then is
equidistantly divided into $4N^2$ small squares, $h = (2N)^{-1}$
is the mesh-size. ϕ_h is determined by its values at the grid
points, it may be identified with its grid function:

$$\phi_h \in C_h = \text{set of grid functions .} \qquad (6)$$

A coarse grid of mesh size $H = 2h$ is introduced, let
$C_H \subset C_h$ be the corresponding set of grid functions. In this
situation the restrictor $\rho : C_h \to C_H$ and the prolongator
$\pi : C_H \to C_h$ of a two-grid method are defined in a most simple
way, the latter one using mean values only.

Our two-grid method consists of an alternating sequence
of just one Gauß-Jacobi iterative step

$$\tilde{\phi}_h = K_h \phi_h^{(k)} + f_h \qquad (7)$$

on the fine grid, restricting the defect

$$d_h = (I_h - K_h)\tilde{\phi}_h - f_h \qquad (8)$$

on the coarse grid : $d_H = \rho d_h$, and solving exactly the
simplified defect equations

$$(I_H - \rho K_h \pi)\delta_H = d_H ; \qquad (9)$$

finally a new approximate surface source $\phi_h^{(k+1)}$ is
obtained from

$$\phi_h^{(k+1)} = \tilde{\phi}_h - \pi \delta_H . \qquad (10)$$

Using some methods of Hackbusch [4] and Novak [1o]
we are able to show that this procedure may be carried out
and yields a sequence $(\phi_h^{(k)})$ which geometrically converges
to the solution ϕ_h of (5). The same holds for the
corresponding approximate flow fields $(u_h^{(k)})$ asymptotically
in the whole flow region ($\to$[8]). Note that in contrast to the
classical iterative procedures it turns out that the convergence factor is improved when the grid is refined.

SOME NUMERICAL RESULTS

At first we study the influence of the free parameter η
in the ansatz (2) on the numerical accuracy. Here the system
(5) is solved with Gauß' elimination on the coarser grids

with mesh-sizes h = 1/8 (N=4) or h = 1/12 (N=6) resp.
In a test example with known solution (cf. [6]) the strong
influence of η turns out from Fig. 1 , and we may regard η
as a "conditioning parameter".

The efficiency of the two-grid method is studied on a
fine grid with mesh-size h = 1/16 (N=8). The nonsparse
matrix here is of size 867×867 , and the calculations are
performed with simple precision. In this case the corresponding coarse-grid matrix is of size 243×243 , and Gauß'
elimination of the defect equations takes only 1/45 of
that expense with the big matrix. A detailed study yields
that our two-grid method is much less expensive than Gauß-
-Jacobi iteration to obtain a given accuracy - the original
Gauß' elimination is entirely out of range.

Hence in the Tables 1 - 2 we compare both iterative
procedures with different starting values. As a measure for
the accuracy the maximum modulus of the defects
$(I_h - K_h)\psi_h - f_h$ (of any approximate surface source ψ_h) on
the fine grid is taken (as in [12],[14]). In Table 1 we start
with $\phi_h^{(o)} \equiv 0$; it turns out that generally one two-grid
cycle gives the same accuracy as three Gauß-Jacobi iterative
steps. Moreover, in several test series the tendency of
Gauß-Jacobi iteration to slow convergence has been observed.
An artificial but impressive example is that of Table 2,
where we start with $\phi_h^{(o)} \equiv 10^6$. The two-grid method works
as usual but Gauß-Jacobi proves to be insufficient.

In the Tables 1 - 2 we abbreviate .οοο45 by .45 -3,eg.

Figure 1

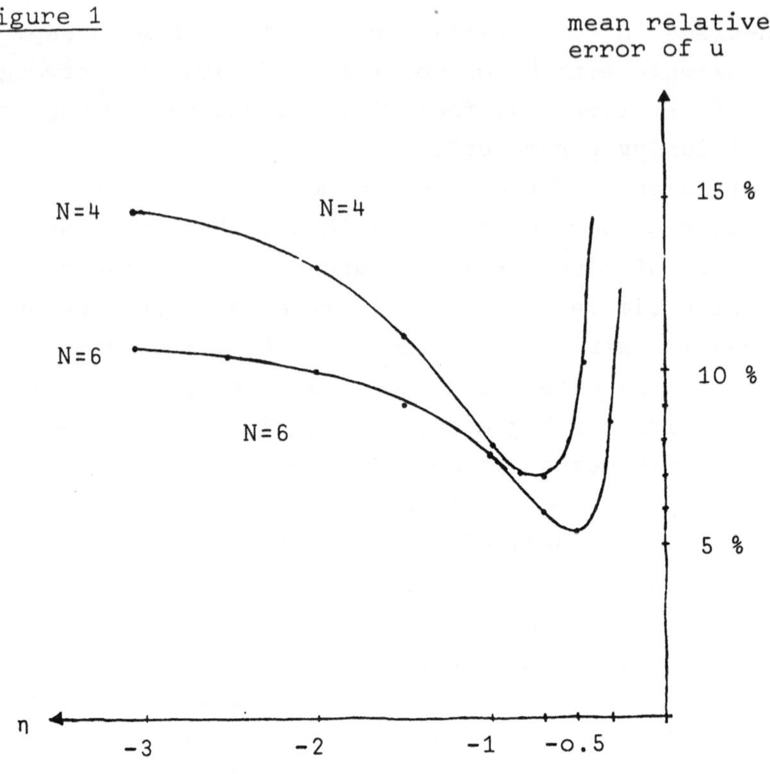

Table 1

no. of cycle	two-grid method	Gauß-Jacobi
1	.24	1.5
2	.38 −1	.76
3	.69 −2	.38
4	.77 −3	.19
5	.1o −3	.1o
6	.13 −4	.5o −1
7	.2o −5	.25 −1
⋮	⋮	⋮
15		.11 −3
⋮		⋮
2o		.4o −5

Table 2

no. of cycle	two-grid method	Gauß-Jacobi
1	19ooo.	77ooo.
2	13oo.	12ooo.
3	99.	28oo.
4	8.	76o.
5	.58	21o.
6	.6o −1	7o.
7	.65 −2	31.
8	.71 −3	2o.
9	.83 −4	17.
1o	.11 −4	16.
⋮	⋮	⋮
15		12.
⋮		⋮
2o		1o.

REFERENCES

[1] Brebbia, C.A., et al., eds.: *"Boundary element methods"*, Berlin 1983.

[2] Faxén, H.: "Fredholmsche Integralgleichungen zu der Hydrodynamik zäher Flüssigkeiten", Ark.Mat.Astr.Fys. 21A, 14 (1929) 1-40.

[3] Fischer, T.M.: "An integral equation procedure for the exterior 3-D slow viscous flow", Integ.Equ.Oper.Th. 5 (1982) 490-505.

[4] Hackbusch, W.: "Die schnelle Auflösung der Fredholmschen Integralgleichung zweiter Art", Beitr.Num.Math. 9 (1981) 47-62.

[5] Hackbusch, W.; Trottenberg, U., eds.: *"Multigrid methods"* Berlin 1982.

[6] Hebeker, F.K.: "A theorem of Faxén and the boundary integral method for 3-D viscous incompressible fluid flows", preprint, Univ. Paderborn 1982.

[7] Hebeker, F.K.: "A boundary integral approach to compute the 3-D Oseen's flow past a moving body", in: Proc. of the Fifth GAMM- Conference on Numerical Methods in Fluid Mechanics (M. Pandolfi, R. Piva, eds.), Vieweg Verlag, Braunschweig/Wiesbaden, 1984, Vol. 7 of Notes on Numerical Fluid Mechanics.

[8] Hebeker, F.K.: "Zur Randelemente-Methode in der 3-D viskosen Strömungsmechanik", Univ. Paderborn, to appear.

[9] Ladyzhenskaja, O.A.: *"The mathematical theory of viscous incompressible fluid flow"*, New York 1969.

[10] Novak, Z.P.: "Use of the multigrid method for Laplacean problems in 3-D", in: [5] , 576-598.

[11] Ries,M.; Trottenberg, U.; Winter, G.: "A note on MGR methods", Lin.Alg.Appl. 49 (1983) 1-26.

[12] Schippers, H.: "Application of multigrid methods for integral equations to two problems from fluid dynamics", J.Comp.Phys. 48 (1982) 441-461.

[13] Wendland, W.: "Die Behandlung von Randwertaufgaben im $\mathbb{R}^3$ mit Hilfe von Einfach- und Doppelschichtpotentialen" Num.Math. 11 (1968) 380-404.

[14] Wolff, H.: "Multiple grid method for the calculation of potential flow around 3-D bodies", preprint, Mathematisch Centrum Amsterdam 1982.

[15] Zhu, J.: "A boundary integral equation method for the stationary Stokes' problem in 3-D", in [1],283-292.

CONFORMING FE-METHOD FOR OBTAINING THE GRADIENT OF A SOLUTION TO THE POISSON EQUATION

P. Neittaanmäki [1]   and   M. Křížek [2]

[1] Lappeenranta University of Technology, Department of Physics and Mathematics, Box 20, SF-53851 Lappeenranta, Finland

[2] Mathematical Institute, Czechoslovak Academy of Sciences, Žitná 25, CS-11567 Prague 1, Czechoslovakia

SUMMARY

We examine a finite element method for the numerical approximation of the gradient of a solution to the Poisson equation. The key is that the gradient appears as a computational variable in a FE-approximation of some first order system. Piecewise linear element fields are used and their approximation properties are studied in the 2- and 3-dimensional case. Numerical examples indicating the accuracy of the method are given.

1. INTRODUCTION

Many mathematical models of physical phenomena lead to the problem

$$\Delta z = f \in L^2(\Omega) \ ,$$
$$z = 0 \quad \text{on} \quad \partial\Omega \ , \tag{1.1}$$

where $\Omega \subset \mathbb{R}^d$ ($d = 2,3$) is a bounded domain. The gradient of z is often more important than the solution z itself. Thus many FE-methods have been developed to calculate this gradient. Let us mention for instance equilibrium methods based on the variational principle of minimum complementary energy [9,11], mixed methods based on the Hellinger-Reissner principle [15] or hybrid methods (for a survey of these methods see [4], Chapter 7).

Consider the problem

$$\begin{aligned}
\operatorname{div} u &= f &&\text{in } \Omega \ , \\
\operatorname{rot} u &= g &&\text{in } \Omega \ , \\
n \wedge u &= 0 &&\text{on } \partial\Omega \ .
\end{aligned} \qquad (1.2)$$

Setting $u = \operatorname{grad} z$, where z solves (1.1), we see that (1.2) holds for $g = 0$. The system (1.2) describes e.g. stress fields in continuum mechanics, electromagnetic fields (the stationary or time harmonic case), velocity fields in the ideal fluid flow problem [3,4,5,14,15,16]. For solving (1.2), the Galerkin method based on the Kelvin principle is proposed in [6]. The least square FE-approximation of (1.2) for smooth domains is studied in [17,18,21]. The aim of the present paper is to generalize some results of [1,3,5,17,18] to non-smooth domains and also some results of [13,14,18] to the three-dimensional case.

Let us still note that the least square FE-approximation requires a certain regularity of the exact solution. On the other hand, when Ω is convex, we get an approximation not only of $\operatorname{grad} z$, but also of all second derivatives of z and we use only C^0-elements.

2. SOME FUNCTION SPACES

Throughout the paper, $\Omega \subset \mathbf{R}^d$, $d = 2 , 3$, will be always a bounded domain with a Lipschitz boundary $\partial\Omega$ (see [15], p. 17) with the outward unit normal $n = (n_1 , \ldots , n_d)$. Notations $H^k(\Omega)$, $k = 0 , 1 , \ldots$, are used for the Sobolev spaces. The usual norm in $H^k(\Omega)$ and also in $(H^k(\Omega))^p$, $p = 2 , 3$, will be denoted by $\|\cdot\|_k$. We denote by $(\cdot , \cdot)_0$ the scalar product in $(L^2(\Omega))^p$, $p = 1 , 2 , 3$. Further, $H^{1/2}(\partial\Omega)$ is the space of traces of functions from $H^1(\Omega)$, and $H^1_0(\Omega)$ is the subspace of $H^1(\Omega)$, consisting of functions with zero traces. Denote by $\mathcal{D}(\Omega)$ the space of infinitely differentiable functions with a compact support in Ω . Let us introduce the following operators

$$\operatorname{curl} \varphi = (\partial_2 \varphi , -\partial_1 \varphi) \qquad \text{for } \varphi \in H^1(\Omega) \ , \quad d = 2 \ ,$$

$$\operatorname{curl} \varphi = (\partial_2 \varphi_3 - \partial_3 \varphi_2 , \ \partial_3 \varphi_1 - \partial_1 \varphi_3 , \ \partial_1 \varphi_2 - \partial_2 \varphi_1)$$
$$\text{for } \varphi = (\varphi_1 , \varphi_2 , \varphi_3) \in (H^1(\Omega))^3 \ , \quad d = 3$$

and the spaces

$$H(\text{div}; \Omega) = \left\{ v \in (L^2(\Omega))^d \mid \exists \bar{f} \in L^2(\Omega) : (v, \text{grad } \varphi)_0 + (\bar{f}, \varphi)_0 = 0 \right.$$
$$\left. \forall \varphi \in \mathcal{D}(\Omega) \right\},$$

$$H(\text{rot}; \Omega) = \left\{ v \in (L^2(\Omega))^d \mid \exists \bar{g} \in (L^2(\Omega))^{2d-3} : (v, \text{curl } \varphi)_0 = (\bar{g}, \varphi)_0 \right.$$
$$\left. \forall \varphi \in (\mathcal{D}(\Omega))^{2d-3} \right\}.$$

Here the functions $\bar{f}$ and $\bar{g}$ are called the <u>divergence</u> and <u>rotation</u> of v, respectively. For $v \in (H^1(\Omega))^d$ it is easy to show that $\text{rot } v = \partial_1 v_2 - \partial_2 v_1$ if $d = 2$, and $\text{rot } v = \text{curl } v$ if $d = 3$ (i.e. the operators curl and rot are identical for $d = 3$).

We recall (see [8], p. 16) that the functional $v \mapsto n \cdot v \big|_{\partial\Omega}$ defined on $(C^\infty(\bar{\Omega}))^d$ can be extended by continuity to a linear continuous mapping from the space $H(\text{div}; \Omega)$ into $H^{-1/2}(\partial\Omega)$, the latter being the dual space to $H^{1/2}(\partial\Omega)$. Analogously for $v \in H(\text{rot}; \Omega)$, the functional $n \wedge v \in (H^{-1/2}(\partial\Omega))^{2d-3}$, can be defined (see also [8]) such that if $v \in (H^1(\Omega))^d$ then $n \wedge v = n_1 v_2 - n_2 v_1$ for $d = 2$, and $n \wedge v$ is the usual vector product for $d = 3$.

Now the Green formulas can be rewritten as

$$(\text{div } v, z)_0 + (v, \text{grad } z)_0$$
$$= \langle n \cdot v, z \rangle_{\partial\Omega} \quad \forall v \in H(\text{div}; \Omega) \quad \forall z \in H^1(\Omega), \tag{2.1}$$

$$(\text{rot } v, z)_0 - (v, \text{curl } z)_0$$
$$= \langle n \wedge v, z \rangle_{\partial\Omega} \quad \forall v \in H(\text{rot}; \Omega) \quad \forall z \in H^1(\Omega))^{2d-3}, \tag{2.2}$$

where $\langle \cdot, \cdot \rangle_{\partial\Omega}$ denotes the duality pairing between $(H^{-1/2}(\partial\Omega))^p$ and $(H^{1/2}(\partial\Omega))^p$ for $p = 1$ or 3.

Now, we define several subspaces of $H(\text{div}; \Omega)$ and $H(\text{rot}; \Omega)$,

$$H_0(\text{div}; \Omega) = \left\{ v \in H(\text{div}; \Omega) \mid n \cdot v = 0 \text{ on } \partial\Omega \right\},$$

$$H(\text{div}^0; \Omega) = \left\{ v \in H(\text{div}; \Omega) \mid \text{div } v = 0 \text{ in } \Omega \right\},$$

$$H_0(\text{div}^0; \Omega) = H_0(\text{div}; \Omega) \cap H(\text{div}^0; \Omega),$$

$$H_0(\text{rot}; \Omega) = \left\{ v \in H(\text{rot}; \Omega) \mid n \wedge v = 0 \text{ on } \partial\Omega \right\},$$

$$H(\text{rot}^0; \Omega) = \left\{ v \in H(\text{rot}; \Omega) \mid \text{rot } v = 0 \text{ in } \Omega \right\},$$

$$H_0(\text{rot}^0; \Omega) = H_0(\text{rot}; \Omega) \cap H(\text{rot}^0; \Omega),$$

$$H_D = H_0(\text{div}^0; \Omega) \cap H(\text{rot}^0; \Omega),$$

$$H_R = H(\text{div}^0 \, ; \, \Omega) \, \cap \, H_0(\text{rot}^0 \, ; \, \Omega) \, ,$$

$$V = H(\text{div} \, ; \, \Omega) \, \cap \, H_0(\text{rot} \, ; \, \Omega) \, .$$

From the density $\overline{\mathcal{D}(\Omega)} = H_0^1(\Omega)$, (2.1) and (2.2) we easily get that

$$\text{grad } z \in H(\text{rot}^0 \, ; \, \Omega) \qquad \text{for} \quad z \in H^1(\Omega) \, , \tag{2.3}$$

$$\text{grad } z \in H_0(\text{rot}^0 \, ; \, \Omega) \qquad \text{for} \quad z \in H_0^1(\Omega) \, , \tag{2.4}$$

$$\text{curl } v \in H(\text{div}^0 \, ; \, \Omega) \quad \begin{cases} \text{for} \quad v \in H^1(\Omega) \quad \text{and} \quad d = 2 \, , \\ \text{for} \quad v \in H(\text{rot} \, ; \, \Omega) \quad \text{and} \quad d = 3 \, , \end{cases} \tag{2.5}$$

and

$$\text{curl } v \in H_0(\text{div}^0 \, ; \, \Omega) \quad \begin{cases} \text{for} \quad v \in H_0^1(\Omega) \quad \text{and} \quad d = 2 \, , \\ \text{for} \quad v \in H_0(\text{rot} \, ; \, \Omega) \quad \text{and} \quad d = 3 \, . \end{cases} \tag{2.6}$$

3. VARIATIONAL PROBLEM

In this chapter we investigate the solvability of a variational problem: Find $u \in V$ such that

$$b(u \, , \, v) = F(v) \quad \forall \, v \in V \, , \tag{3.1}$$

where F is a linear continuous functional on V and

$$b(u \, , \, v) = (\text{div } u \, , \, \text{div } v)_0 + (\text{rot } u \, , \, \text{rot } v)_0 \, . \tag{3.2}$$

<u>Theorem 3.1.</u> Let $\Omega \subset \mathbf{R}^d$, $d = 2 , 3$, be a bounded domain with a Lipschitz boundary. Then

$$\|v\|_0 \leq C(\|\text{div } v\|_0 + \|\text{rot } v\|_0) \, , \tag{3.3}$$

$$\forall \, v \in V = H(\text{div} \, ; \, \Omega) \, \cap \, H_0(\text{rot} \, ; \, \Omega)$$

if and only if $\partial\Omega$ is connected.

The proof is based on two lemmas.

<u>Lemma 3.2.</u> Let $\Omega \subset \mathbf{R}^3$ be a bounded domain with a Lipschitz boundary. Then

$$(\text{rot } s \, , \, \text{rot } s)_0 = (s \, , \, \text{rot rot } s)_0$$

for all $s \in H(\text{rot} \, ; \, \Omega)$ such that $\text{rot } s \in H_0(\text{rot} \, ; \, \Omega)$.

$\underline{Proof}$. Let $s \in H(\text{rot} ; \Omega)$ with $\text{rot } s \in H_0(\text{rot} ; \Omega)$ be given. Since $(\mathcal{D}(\Omega))^3$ is dense in $H_0(\text{rot} ; \Omega)$ with respect to the norm:

$$\| \cdot \|_{H(\text{rot};\Omega)} = (\| \cdot \|_0^2 + \|\text{rot } \cdot \|_0^2)^{1/2}$$

(see [8], p. 21), there exists a sequence $\varphi_j \in (\mathcal{D}(\Omega))^3$ such that

$$\|\text{rot } s - \varphi_j\|_{H(\text{rot};\Omega)} \to 0 \qquad \text{for} \quad j \to \infty .$$

Thus, we conclude that

$$(\text{rot } s , \varphi_j)_0 \to (\text{rot } s , \text{rot } s)_0 \tag{3.4}$$

and

$$(s , \text{rot } \varphi_j)_0 \to (s , \text{rot rot } s)_0 \tag{3.5}$$

when $j \to \infty$. By the Green formula (2.2), we come to

$$(\text{rot } s , \varphi_j)_0 - (s , \text{rot } \varphi_j)_0 = \langle n \wedge s , \varphi_j \rangle = 0 .$$

This together with (3.4) and (3.5) yields the assertion of the lemma. $\square$

Let us denote by $X^\perp$ the orthocomplement of a closed subspace $X \subset (L^2(\Omega))^3$ in $(L^2(\Omega))^3$.

$\underline{\text{Lemma 3.3.}}$ Let $\Omega \subset \mathbb{R}^3$ be a bounded domain with a connected Lipschitz boundary and let $w \in H(\text{div}^0 ; \Omega)$ be given. Then there exists unique $s \in S = Q \cap (H_D)^\perp$, $Q = H_0(\text{div}^0 ; \Omega) \cap H(\text{rot} ; \Omega)$, such that

$$w = \text{rot } s ,$$

and

$$\|s\|_0 \leq C \|\text{rot } s\|_0 ,$$

where $C > 0$ does not depend on s and w .

$\underline{\text{Proof}}$. Let the assumptions be satisfied. By [8], p. 28, there exists the so-called stream function $q' \in H(\text{div}^0 ; \Omega) \cap (H^1(\Omega))^3$ (not uniquely determined) such that $w = \text{rot } q'$. As $\langle n \cdot q' , 1 \rangle_{\partial\Omega} = 0$ due to (2.1), the following Neumann problem is solvable

$$\begin{cases} \Delta\psi = 0 & \text{in } \Omega , \\[2mm] \dfrac{\partial\psi}{\partial n} = n \cdot q' & \text{on } \partial\Omega . \end{cases}$$

Then clearly the function $q = q' - \text{grad } \psi$ is from Q and q is also a stream function to w , that is

$$w = \text{rot } q \, ,$$

(cf. [2,3,22]). From the orthogonal decomposition

$$Q = S \oplus H_D \, , \tag{3.6}$$

we get $q = s + h$, where $s \in S$ and $h \in H_D$. Hence, from the definition of H_D we obtain $w = \text{rot } s$.

Further, let us suppose that $w = \text{rot } s^1 = \text{rot } s^2$ for some $s^1 , s^2 \in S$. Then $s^1 - s^2 \in S \cap H_D$ and (3.6) yields $s^1 - s^2 = 0$, i.e. the stream function $s \in S$ exists and is unique. Therefore, from (2.5) we obtain that the linear operator

$$\text{rot: } S \to H(\text{div}^0 ; \Omega) \tag{3.7}$$

is bijective. It is obvious that S with the $\| \cdot \|_{H(\text{rot};\Omega)}$-norm and $H(\text{div}^0 ; \Omega)$ with the $\| \cdot \|_0$-norm are Banach spaces. As the operator (3.7) is continuous, that is

$$\| \text{rot } s \|_0 \leq C' \ \| s \|_{H(\text{rot};\Omega)} \, ,$$

we conclude by the closed graph theorem that the inverse operator is continuous, too, and it holds that

$$\| s \|_0 \leq \| s \|_{H(\text{rot};\Omega)} \leq C \ \| \text{rot } s \|_0 \, . \quad \square$$

<u>Proof of Theorem 3.1.</u> Let $\partial\Omega$ be not connected and let Γ be one of its components. Consider the problem

$$\begin{cases} \Delta z \doteq 0 & \text{in } \Omega \, , \\ z = 1 & \text{on } \Gamma \, , \\ z = 0 & \text{on } \partial\Omega - \Gamma \, . \end{cases}$$

We easily find that $v = \text{grad } z \in H_R = H(\text{div}^0 ; \Omega) \cap H_0(\text{rot}^0 ; \Omega)$, while $\| v \|_0 \neq 0$, i.e. (3.3) cannot hold.

Conversely, let $\partial\Omega$ be connected, let $v \in V$ be given and let $d = 3$. The case $d = 2$ is proved in [13], Theorem 4.4. Let $z \in H_0^1(\Omega)$ be a weak solution of the problem

$$\begin{cases} \Delta z = \text{div } v & \text{in } \Omega \, , \\ z = 0 & \text{on } \partial\Omega \, . \end{cases} \tag{3.8}$$

80

Thus it holds that

$$\|z\|_1 \leq c_1 \|\text{div } v\|_0 . \qquad (3.9)$$

From the weak formulation of the problem (3.8), from (2.1), and (2.4) we see that $\text{grad } z \in H(\text{div}; \Omega) \cap H_0(\text{rot}^0; \Omega)$, i.e. by (3.8) we get

$$w = v - \text{grad } z \in H(\text{div}^0; \Omega) \cap H_0(\text{rot}; \Omega) . \qquad (3.10)$$

According to Lemma 3.3 there exists a stream function $s \in S \subset H(\text{rot}; \Omega)$ such that

$$w = \text{rot } s \in H_0(\text{rot}; \Omega) . \qquad (3.11)$$

Applying now Lemmas 3.2 and 3.3, we come to

$$\|\text{rot } s\|_0^2 = (\text{rot } s , \text{rot } s)_0 = (s , \text{rot rot } s)_0$$
$$\leq \|s\|_0 \|\text{rot rot } s\|_0 \leq c_2 \|\text{rot } s\|_0 \|\text{rot rot } s\|_0 . \qquad (3.12)$$

Thus by (3.10), (3.11), (3.12), (3.9), and by (2.3) we obtain

$$\|v\|_0 \leq \|\text{grad } z\|_0 + \|\text{rot } s\|_0 \leq \|z\|_1 + c_2 \|\text{rot rot } s\|_0$$
$$\leq c_1 \|\text{div } v\|_0 + c_2 \|\text{rot } w\|_0 \leq c(\|\text{div } v\|_0 + \|\text{rot } v\|_0) . \quad \square$$

<u>Remark 3.4.</u> For a bounded convex domain or for a smooth domain, which is homeomorphic to a ball, it even holds that (see [3,7,13,20])

$$\|v\|_1 \leq c(\|\text{div } v\|_0 + \|\text{rot } v\|_0) \quad \forall \, v \in V . \qquad (3.13)$$

<u>Remark 3.5.</u> The spaces H_D and H_R are finite-dimensional (see [19]). From Theorem 3.1 we see that H_R is trivial iff $\partial\Omega$ is connected; (note that H_D is trivial iff Ω is simply connected, see [1]). The proof of the inequality (3.3) can be easily modified for $v \in V \cap (H_R)^\perp$ without any assumptions on the connectivity of $\partial\Omega$.

However, for simplicity <u>we shall assume that $\partial\Omega$ is connected in what</u> <u>follows.</u>

<u>Theorem 3.6.</u> The problem (3.1) has exactly one solution.

<u>Proof.</u> By Theorem 3.1 the bilinear form (3.2) is a scalar product on V , and it is easy to show that V will be now the Hilbert space. Thus, the assertion follows from the Riesz theorem. $\square$

4. APPLICATION

In this Chapter we shall show that the problem of finding $u = \operatorname{grad} z$, where $z \in H_0^1(\Omega)$ is the solution of (1.1), can be variationally formulated by (3.1).

We immediately see from (1.1) and (2.4) that for $u = \operatorname{grad} z$ we have $u \in V$ and

$$\operatorname{div} u = f \ , \tag{4.1}$$

$$\operatorname{rot} u = 0 \ . \tag{4.2}$$

Hence, u satisfies (3.1) for

$$F(v) = (f , \operatorname{div} v)_0 \ , \quad v \in V \ . \tag{4.3}$$

Conversely, let $u \in V$ fulfil (3.1) with F defined by (4.3). We prove that $u = \operatorname{grad} z$ for z from (1.1). First, we verify that (4.1) and (4.2) are satisfied. Let $\varphi \in L^2(\Omega)$ be arbitrary and let $v = \operatorname{grad} \chi$, where $\chi \in H^1(\Omega)$ is a weak solution of the Dirichlet problem

$$\begin{cases} \Delta\chi = \varphi & \text{in } \Omega \ , \\ \chi = 0 & \text{on } \partial\Omega \ . \end{cases}$$

As $v \in V$ due to (2.4), we obtain from (3.1), (3.2), and (4.3) that $(\operatorname{div} u , \varphi)_0 = (\operatorname{div} u , \operatorname{div} v)_0 = b(u , v) = F(v) = (f , \operatorname{div} v)_0 = (f , \varphi)_0$, i.e. (4.1) holds.

To prove (4.2) we distinguish two cases:

a) $d = 2$. In accordance with the Green formula (2.2) we have

$$(\operatorname{rot} u , 1)_0 = \langle n \wedge u , 1 \rangle_{\partial\Omega} = 0 \ . \tag{4.4}$$

So let $\psi \in L^2(\Omega)$ with

$$(\psi , 1)_0 = 0 \tag{4.5}$$

be arbitrary. Putting $v = \operatorname{curl} \omega$, where $\omega \in H^1(\Omega)$, $(\omega , 1)_0 = 0$, is a weak solution of the Neumann problem

$$-\Delta\omega = \psi \quad \text{in } \Omega \ ,$$
$$\frac{\partial\omega}{\partial n} = 0 \quad \text{on } \partial\Omega \ ,$$

we find that $v \in V$, since $n \wedge \mathrm{curl}\ \omega = - n \cdot \mathrm{grad}\ \omega = 0$ in $H^{-1/2}(\partial\Omega)$, rot curl $\omega = - \Delta\omega$ in $L^2(\Omega)$, and since v is divergence-free. Consequently,

$$(\mathrm{rot}\ u\ ,\ \psi)_0 = (\mathrm{rot}\ u\ ,\ \mathrm{rot}\ v)_0 = b(u\ ,\ v) = F(v) = 0 \ . \qquad (4.6)$$

Now, a combination of (4.4), (4.5), and (4.6) gives (4.2).

b) $d = 3$. Using (2.2) for $h \in H_D$, we get

$$(\mathrm{rot}\ u\ ,\ h)_0 = (u\ ,\ \mathrm{rot}\ h)_0 + <n \wedge u\ ,\ h>_{\partial\Omega} = 0 \ ,$$

i.e. rot $u \in (H_D)^\perp$, which together with (2.6) yields

$$\mathrm{rot}\ u \in H_0(\mathrm{div}^0\ ;\ \Omega) \cap (H_D)^\perp \ . \qquad (4.7)$$

Let us arbitrarily choose $\psi \in H_0(\mathrm{div}^0\ ;\ \Omega) \cap (H_D)^\perp$. Then there exists a stream function $v \in H(\mathrm{div}^0\ ;\ \Omega) \cap H_0(\mathrm{rot}\ ;\ \Omega) \subset V$ such that $\psi = \mathrm{rot}\ v$ (see e.g. [2,3,22]). Similarly as we got (4.6) we prove that $(\mathrm{rot}\ u\ ,\ \psi)_0 = 0$, so (4.2) holds again.

Now the relations (1.1), (2.4), (4.1), and (4.2) imply that $u - \mathrm{grad}\ z \in V$ and

$$\mathrm{div}(u - \mathrm{grad}\ z) = 0 \ ,$$

$$\mathrm{rot}(u - \mathrm{grad}\ z) = 0 \ .$$

Hence, $u = \mathrm{grad}\ z$, which follows from Theorem 3.1.

<u>Remark 4.1.</u> The problem (1.2) can be variationally formulated by (3.1) for

$$F(v) = (f\ ,\ \mathrm{div}\ v)_0 + (g\ ,\ \mathrm{rot}\ v)_0 \ , \qquad v \in V \ .$$

5. FINITE ELEMENT APPROXIMATION

To obtain a conforming FE-method we shall suppose that Ω is polygonal or polyhedral, (for curved boundaries see [14,17,18,21]). Let $\{T_h\}$ be a regular family of decompositions of $\bar\Omega$ into triangles or tetrahedra (for the existence of such a family in case $d = 3$ see [12], p. 58). Let

$$V_h = \left\{ v_h \in (C(\bar\Omega))^d \mid v_h|_K \in (P_1(K))^d\ \forall K \in T_h\ ,\ n \wedge v_h = 0 \ \text{on}\ \partial\Omega \right\} \ ,$$

where $P_1(K)$ is the space of linear polynomials on K .

The discrete problem of (3.1) will consist in finding $u_h \in V_h$ such that

$$b(u_h , v_h) = F(v_h) \quad \forall \ v_h \in V_h \ . \tag{5.1}$$

This problem has just one solution since clearly $V_h \subset V$ (the so-called conformity condition). By Theorem 3.1 the norm

$$|||v||| = (||v||_0^2 + ||\text{div } v||_0^2 + ||\text{rot } v||_0^2)^{1/2} \ , \qquad v \in V \ , \tag{5.2}$$

is equivalent with $(b(v , v))^{1/2}$, and we have the following theorems.

<u>Theorem 5.1.</u> Let $u \in (H^2(\Omega))^d$ be the solution of (3.1). Then

$$|||u - u_h||| \leq C \ h \ ||u||_2 \ . \tag{5.3}$$

<u>Proof</u>. Applying Cea's Lemma ([4], p. 104), we obtain

$$|||u - u_h||| \leq C_1 \inf_{v_h \in V_h} |||u - v_h||| \leq C_1 \inf_{v_h \in V_h} ||u - v_h||_1 \leq C_2 \ h \ ||u||_2 \ ,$$

where the last estimate follows taking v_h as the V_h-interpolant of u . $\square$

<u>Theorem 5.2.</u> Let Ω be polygonal and let $u \in (H^1(\Omega))^2$. Then

$$|||u - u_h||| \to 0 \qquad \text{as} \quad h \to 0 \ . \tag{5.4}$$

<u>Proof</u>. By [10], Theorem 3.1, the space $V \cap (H^1(\Omega))^2$ contains a dense subset of infinitely differentiable functions with respect to the $||\cdot||_1$-norm. Thus for a given $\varepsilon > 0$ there exists $w \in V \cap (C^\infty(\bar{\Omega}))^2$ such that $||u - w||_1 < \varepsilon/2$. Hence,

$$|||u - u_h||| \leq |||u - w||| + |||w - u_h||| \leq ||u - w||_1 + C_1 \inf_{v_h \in V_h} ||w - v_h||_1$$

$$= \varepsilon/2 + \varepsilon/2 = \varepsilon \ . \ \square$$

<u>Remark 5.3.</u> Using Theorem 5.2 and (5.2), we find that

$$u_h \to \text{grad } z \ , \tag{5.5}$$

$$\text{div } u_h \to \Delta z \ , \tag{5.6}$$

and

$$\text{rot } u_h \to 0 \tag{5.7}$$

in the $||\cdot||_0$-norm if the solution z of (1.1) belongs to $H^2(\Omega)$. Moreover, for $z \in H^3(\Omega)$ Theorem 5.1 yields

$$\| \Delta z - \mathrm{div}\, u_h \|_0 \leq C\, h\, \| z \|_3 \, . \tag{5.8}$$

Remark 5.4. From (3.13) we see that the $\|\|\cdot\|\|$-norm is equivalent to the $\|\cdot\|_1$-norm for a convex domain Ω . Therefore, the solution u ($=\mathrm{grad}\, z$) of (3.1) is in this case always from $(H^1(\Omega))^d$. Consequently, we get by the presented FE-method approximation even of all second derivatives of z .

6. NUMERICAL TESTS

The method given above was tested in different geometries - see [14]. In the following we present two examples on the unit square.

Example 6.1. Let $\Omega = (0\,,\,1) \times (0\,,\,1)$. For

$$f(x_1\,,\,x_2) = -2 \sin \pi x_2 - \pi^2 x_1 (1-x_1) \sin \pi x_2 \,,$$

the corresponding solution of (1.1) is

$$z(x_1\,,\,x_2) = x_1 (1-x_1) \sin \pi x_2 \, .$$

The approximation of $u = \mathrm{grad}\, z$ was computed from (5.1). The values of the error $u - u_h$ in various norms are shown in Table 6.1.

h^{-1}	$\| u - u_h \|_0$	$\|\| u - u_h \|\|$	$\| u - u_h \|_1$	$\| \Delta z - \mathrm{div}\, u_h \|_0$
2	0.14718	2.08084	3.71981	1.82542
4	0.03882	1.28104	2.08152	1.09224
8	0.00969	0.67642	1.07013	0.57023
16	0.00242	0.34299	0.53784	0.28817

Table 6.1.

From this table it can be seen that the convergence in the $\|\cdot\|_0$-norm is quadratic (cf. [10]) whereas it is linear in the $\|\|\cdot\|\|$-norm or $\|\cdot\|_1$-norm (see Theorem 5.1 and (3.13)). The last column confirms the linear convergence stated in (5.8).

Example 6.2. Let $\Omega = (0\,,\,1) \times (0\,,\,1)$ and let $f(x_1\,,\,x_2) = -2$ in (1.1). Such a problem describes the torsion of an isotropic homogeneous elastic bar of square cross section and the solution z is called the Prandtl potential function.

Figure 6.2 shows the finite element grid for $h = 1/16$ as well as the corresponding FE-approximation of grad z .

 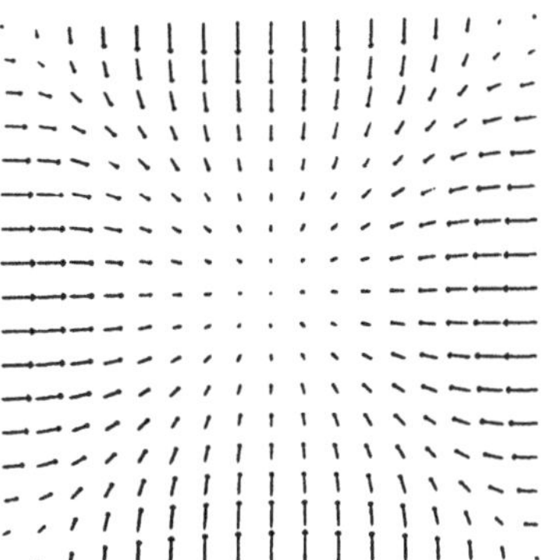

<u>Figure 6.2.</u> Triangulation $T_{1/16}$ of $\bar{\Omega}$ and corresponding FE-approximation for grad z .

Note that grad z is proportional to the vector $(-\tau_{23}, \tau_{13})$, where τ_{ij} are the components of the symmetric stress tensor τ in the elastic bar ($\tau_{11} = \tau_{22} = \tau_{33} = \tau_{12} = 0$). For more details we refer to [15], p. 229.

REFERENCES

[1] Bendali, A., Dominguez, J.M., and Gallic, S.: <u>A variational approach for the vector potential formulation of the Stokes and Navier-Stokes problems in three dimensional domains</u>; Ecole Polytechnique, Rapport interne no 95 (1983), pp 1-29.

[2] Bernardi, C.: <u>Formulation variationnelle mixte des equations de Navier-Stokes en dimension 3</u>; Thèse de 3ème cycle (deuxieme partie), Paris VI (1979), pp 146-176.

[3] Byhovskiĭ, È.B.: <u>Solution of a mixed problem for the system of Maxwell equations in case of ideally conductive boundary</u> (Russian); Vestnik Leningrad. Univ. Mat. Meh. Astronom. 12 (1957), no 13, pp 50-66.

[4] Ciarlet, P.G.: <u>The finite element method for elliptic problems</u>; North-Holland Publishing Company, Amsterdam, New York, Oxford (1978).

[5] Duvaut, G., and Lions, J.L.: <u>Inequalities in mechanics and physics</u>; Springer-Verlag, Berlin, Heidelberg, New York (1976).

[6] Fix, G.J., Gunzburger, M.D., and Nicolaides, R.A.: <u>On mixed finite element methods for first order elliptic systems</u>; Numer. Math. 37 (1981), pp 29-48.

[7] Georgescu, V.: Some boundary value problem for differential forms on compact Riemannian manifolds; Ann. Mat. Pura Appl., Série IV, 122 (1979), pp 159-198.

[8] Girault, V., and Raviart, P.A.: Finite element approximation of the Navier-Stokes equation; Springer-Verlag, Berlin, Heidelberg, New York (1979).

[9] Haslinger, J., and Hlaváček, I.: Convergence of a finite element method based on the dual variational formulation; Apl. Mat. 21 (1976), pp 43-65.

[10] Haslinger, J., and Neittaanmäki, P.: On different finite element methods for approximating the gradient of the solution to the Helmholtz equation; Comput. Methods Appl. Mech. Engrg. 42 (1984), pp 131-148.

[11] Hlaváček, I., and Křížek, M.: Internal finite element approximation in the dual variational methods for second order elliptic problems with curved boundaries; Apl. Mat. 29 (1984), pp 52-69.

[12] Křížek, M.: An equilibrium finite element method in three-dimensional elasticity; Apl. Mat. 27 (1982), pp 46-75.

[13] Křížek, M., and Neittaanmäki, P.: On the validity of Friedrichs' inequalities; Math. Scand. 54 (1984).

[14] Křížek, M., and Neittaanmäki, P.: Finite element approximation for a div-rot system with mixed boundary conditions in non-smooth plane domains; Apl. Mat. 29 (1984), to appear.

[15] Nečas, J., and Hlaváček, I.: Mathematical theory of elastic and elasto-plastic bodies: an introduction; Elsevier Scientific Publishing Company, Amsterdam, Oxford, New York (1981).

[16] Neittaanmäki, P., and Picard, R.: Error estimates for the finite element approximation to a Maxwell-type boundary value problem; Numer. Funct. Anal. Optim. 2 (1980), pp 267-285.

[17] Neittaanmäki, P., and Saranen, J.: On the finite element approximation of gradient for solution of Poisson equation; Numer. Math. 37 (1981), pp 333-337.

[18] Neittaanmäki, P., and Saranen, J.: Finite element approximation of electromagnetic fields in the three dimensional space; Numer. Funct. Anal. Optim. 2 (1981), pp 487-506.

[19] Picard, R.: An elementary proof for a compact imbedding result in generalized electromagnetic theory; SFB 72, Universität Bonn, preprint 624 (1983).

[20] Saranen, J.: On an inequality of Friedrichs; Math. Scand. 51 (1982), pp 310-322.

[21] Saranen, J.: A least squares approximation method for first-order elliptic systems of plane; Applicable Anal. 14 (1982), pp 27-42.

[22] Weber, Ch.: A local compactness theorem for Maxwell's equations; Math. Methods Appl. Sci. 2 (1980), pp 12-25.

A MULTIGRID-SOLVER FOR THE COMPUTATION OF IN-CYLINDER TURBULENT FLOWS IN ENGINES

Barbara Ruttmann
Karl Solchenbach
Gesellschaft für Mathematik und Datenverarbeitung
Postfach 1240
D-5205 St. Augustin

SUMMARY

In the last 20 years in automobile engine design many interesting applications of computational fluid dynamics (CFD) techniques have arisen. One of them is the numerical simulation of in-cylinder flow and combustion processes in reciprocating engines. The mathematical model for the gas motion in the combustion chamber is the time-dependent, compressible Navier-Stokes equations combined with a standard turbulence model.

This paper presents the multigrid (MG)-program MGNS2O for the efficient numerical solution of the 2D-model without combustion. The time discretization is done implicitly. Therefore, the time-step size is not restricted by a stability condition. The non-linear elliptic system arising in each time-step is solved by a specially designed MG-method using relaxation techniques proposed by A. Brandt.

1. INTRODUCTION

The development of modern automobile engines must take into account the rapidly increasing energy-supply and ecological problems. Legislative constraints on pollutent emissions, the critical public opinion and considerations of economy by the customers require radical alternatives to existing designs. However, these requirements should not be met at the expense of poorer engine performance.

Engines are very sensitive to the fluid-dynamic processes within the cylinder. For instance, a large amount of turbulence is necessary to guarantee good fuel-air mixing and effective combustion. The flow behaviour depends strongly on the geometrical configuration of intake valve, spark plug, piston bowl etc. In order to improve the engine performance, many manufacturers try to optimize the design of the combustion chamber. Since building and testing of research engines is expensive and time-consuming, the role of numerical simulations as an aid to efficient engine design becomes more and more important.

The mathematical description of the physical processes within the cylinder must reflect the following features
- complex geometric configuration with moving boundaries,
- real 3D instationary flows without symmetries,
- propagation of the flame front with steep spatial gradients of temperature and other quantities,
- dependence of the chemical reactions on the mixing and the turbulent components of the gas flow.

In order to reduce the complexity of the mathematical model, the combustion and chemical reactions are neglected in our application. The gas motion is described mathematically by the instationary, compressible, Reynolds-averaged Navier-Stokes equations combined with a second order turbulence model.

Early numerical methods for the solution of these equations are based on standard CFD techniques like the ICE-(Implicit-Continuous fluid-Eulerian-) method [10]. The ICE-technique treats the viscous terms explicity in time and therefore suffers from the well-known stability limitation of the time-step size which is very restrictive in case of flows with high turbulent viscosity. Computer programs which have been recently developped for 2D- and 3D-simulations apply a fully implicit time discretization without any stability restrictions. The large systems occurring are solved iteratively by ADI-methods [1] or the "Strongly Implicit Procedure" [9].

The aim of our work in this context is the development of an efficient algorithm using implicit time discretization in connection with a MG solution technique. MG-methods are known to yield the fastest algorithms for standard problems as Poisson-like equations [14], [17]. Also for special fluid dynamics problems (incompressible Navier-Stokes equations [13], compressible in-

viscid problems [2], [11]) the superiority of the MG approach
starts to become evident. Our attempt is to maintain the essential advantages of MG such as
- convergence factor far below 1 independent of the number of
  gridpoints,
- geometric flexibility,
- control of the discretization error
in a really complex and practically important application.

To gain first results quickly, we concentrated on the treatment
of the complexity which is inherent in the differential equations.
The geometric complexity and the related grid generation problems
were eliminated by taking a 2D square as computational domain.
This may be a curious choice from the engineer's point of view,
but all MG experience indicates that - in the absence of re-
entrant corners which decrease the regularity properties of the
solution - MG performance does not depend on the shape of the
domain [16].

Section 2 contains the mathematical model in two space dimensions
in primitive variables. In Section 3 we describe the spatial and
time discretization and discuss the related stability questions.
The advantages of the implicit time discretization compared to
the explicit scheme used in [10] are pointed out. The details of
the MG-algorithm are explained in Section 4 and numerical results
are presented in Section 5. Some future aspects concerning the
extension of the algorithm to 3D-simulations are discussed in
Section 6.

This work was stimulated by many ideas and experiences of the MG
research group at the GMD and was supervised by Ulrich Trottenberg.
The Volkswagen AG, Wolfsburg, West Germany supported the project
both financially and in substance.

2. MATHEMATICAL MODEL

The motion of gas in the combustion chamber is assumed to obey
the conservation laws for an isotropic, single-phase fluid continuum. In differential form these are the compressible Navier-Stokes
equations; namely three momentum equations for the averaged velocities, the continuity equation for mass conservation, the energy

equation and an equation of state. The k-ε turbulence model is introduced to describe the effect of the turbulent velocity components. For a detailed derivation of the mathematical model see [8] , [10].

Up to now, nearly all computer programs perform a 2D-simulation, where the computational domain is either a cross-section perpendicular to the cylinder axis (x-y-plane) or a longitudinal section assuming axisymmetry (r-z-plane). In order to separate the geometrical (more or less technical) and the mathematical difficulties of the model, we simplify, in the 2D-model, the computational x-y-domain to a square. Thus all questions concerning the treatment of the boundary conditions, the choice of the coordinate system and the grid generation are temporarily eliminated. The extension to a 3D-MG-program, which is the final aim of our work, requires, of course, a careful discussion of these points (see Section 6).

<u>Governing differential equations</u>

Let x,y denote the independent spatial variables and t the time. The Reynolds-averaged Navier-Stokes equations in nonconservative formulation for the dependent variables ρ (density), u,v (velocity components in x- and y-direction), T (temperature) and p (pressure) are

$$\rho\frac{Du}{Dt} - \frac{\partial}{\partial x}\sigma_{xx} - \frac{\partial}{\partial y}\sigma_{xy} + \frac{\partial P}{\partial x} = 0 \tag{2.1.1}$$

$$\rho\frac{Dv}{Dt} - \frac{\partial}{\partial x}\sigma_{yx} - \frac{\partial}{\partial y}\sigma_{yy} + \frac{\partial P}{\partial y} = 0 \tag{2.1.2}$$

$$\frac{D\rho}{Dt} + \rho(\frac{\partial u}{\partial x} + \frac{\partial v}{\partial y}) = 0 \tag{2.1.3}$$

$$\rho c_v\frac{DT}{Dt} - \frac{\partial}{\partial x}(c_p\mu\frac{\partial T}{\partial x}) - \frac{\partial}{\partial y}(c_p\mu\frac{\partial T}{\partial y}) + P(\frac{\partial u}{\partial x} + \frac{\partial v}{\partial y}) - \tag{2.1.4}$$

$$- \mu B(u,v) = S$$

Here $\frac{D}{Dt}$ denotes the substantive derivative $\frac{\partial}{\partial t} + u\frac{\partial}{\partial x} + v\frac{\partial}{\partial y}$,

$$\sigma_{xx} = 2\mu\frac{\partial u}{\partial x} - \frac{2}{3}\mu(\frac{\partial u}{\partial x} + \frac{\partial v}{\partial y}) , \quad \sigma_{xy} = \sigma_{yx} = \mu(\frac{\partial u}{\partial y} + \frac{\partial v}{\partial x}) ,$$

$$\sigma_{yy} = 2\mu\frac{\partial v}{\partial y} - \frac{2}{3}\mu(\frac{\partial u}{\partial x} + \frac{\partial v}{\partial y}) , \quad P = p + \frac{2}{3}\rho k ,$$

$$B(u,v) = \frac{4}{3}((\frac{\partial u}{\partial x})^2 + (\frac{\partial v}{\partial y})^2 - \frac{\partial u}{\partial x}\frac{\partial v}{\partial y}) + (\frac{\partial u}{\partial y} + \frac{\partial v}{\partial x})^2 .$$

$c_v(T)$ and $c_p(T)$ are the specific heats (of nitrogen), S represents the released chemical energy.

The equation of state is

$$p = R \rho T . \qquad (2.1.5)$$

R is the specific gas constant of nitrogen.

The turbulent effects are described by two further unknowns k (turbulent kinetic energy) and ε (dissipation of k) which satisfy the transport equations

$$\rho \frac{Dk}{Dt} - \frac{\partial}{\partial x}(\mu \frac{\partial k}{\partial x}) - \frac{\partial}{\partial y}(\mu \frac{\partial k}{\partial y}) + \frac{2}{3} \rho k \; (\frac{\partial u}{\partial x} + \frac{\partial v}{\partial y}) -$$
$$- c_\mu \rho \frac{k^2}{\varepsilon} B(u,v) + \rho \varepsilon = 0 \qquad (2.2.1)$$

$$\rho \frac{D\varepsilon}{Dt} - (\frac{\partial}{\partial x}(\mu \frac{\partial \varepsilon}{\partial x}) + \frac{\partial}{\partial y}(\mu \frac{\partial \varepsilon}{\partial y})) / \sigma_\varepsilon - (1-\frac{2}{3}c_{\varepsilon_1}) \rho \varepsilon (\frac{\partial u}{\partial x} + \frac{\partial v}{\partial y}) -$$
$$- c_{\varepsilon_1} c_\mu \rho k \, B(u,v) + c_{\varepsilon_2} \rho \frac{\varepsilon^2}{k} = 0. \qquad (2.2.2)$$

The viscosity is given by

$$\mu = \mu_\ell + \mu_t \text{ with } \mu_\ell = 1.42 \cdot 10^{-5} \; T^{1.5} / (T + 100) \text{ and}$$
$$\mu_t = c_\mu \rho k^2 / \varepsilon. \qquad (2.3)$$

The constants in the k $-$ ε equations are determined empirically
$$c_\mu = 0.09, \; \sigma_\varepsilon = 1.3, \; c_{\varepsilon_1} = 1.44, \; c_{\varepsilon_2} = 1.9.$$

A simpler turbulence model without additional quantities (k=0 in eqs. (2.1)) is expressed by the assumption

$$\mu = a_\mu \rho. \qquad (2.4)$$

<u>Initial and boundary values</u>

The initial time t = 0 corresponds to the physical situation of closed inlet valve just after fresh gas with a certain swirl motion has streamed in. The computation then ranges over the compression and power stroke and stops before the outlet valve opens. This time interval corresponds to a crank angle movement of 360° and lasts e.g. 20 msec if the engine speed is 3000 rpm.

The initial values at t = 0 in our model application are given as
$$\rho = \rho_0 = 0.01 \text{ g/cm}^3, \; T = T_0 = 695 \text{ K}, \; p = p_0 = R\rho_0 T_0,$$
$$k = k_0 = 4.09 \cdot 10^6 \text{ cm}^2/\text{sec}^2, \; \varepsilon = \varepsilon_0 = 8.27 \cdot 10^9 \text{ cm}^2/\text{sec}^3. \qquad (2.5)$$

The inital velocity field is shown in Fig. 1. The maximum

value is 2500 cm/sec.

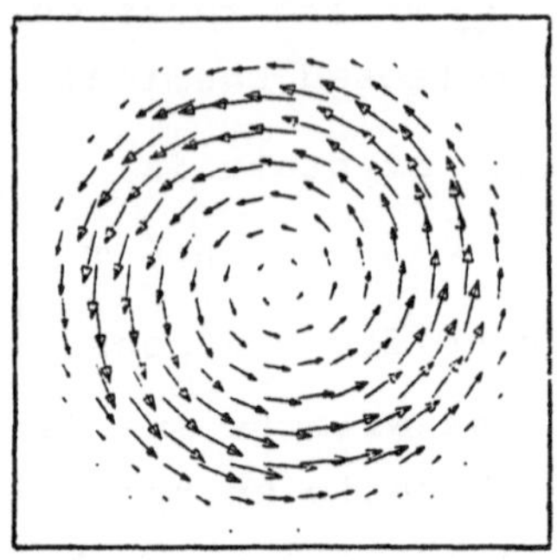

Fig.1: initial velocity field

The size of the square $[0,L]^2$ with L = 8 cm is similar to the area of a real cylinder's cross-section.

At the boundaries we impose the following conditions
u = v = 0 (no-slip condition due to viscosity),

$$\frac{\partial T}{\partial n} = 0 \qquad \text{(no heat transfer through the walls)}, \qquad (2.6)$$

$\frac{\partial k}{\partial n} = 0$, $\varepsilon = \varepsilon(k)$ (resulting from boundary layer
considerations).

The (turbulent) Reynolds- and Mach-Number are considerably lower than in usual aerodynamic applications. If we regard the maximum swirl velocity as reference value u^* and choose

$$\rho^* = \rho_o, \quad \mu^* = c_\mu \rho_o k_o^2 / \varepsilon_o$$

we obtain

$$Re_t = u^* L \rho^* / \mu^* \approx 110,$$

$$M_o = u^* / a_s \approx 0.05 \ (a_s \text{ speed of sound}).$$

3. DISCRETIZATION

Staggered grid

System (2.1-2) is discretized by finite differences using a regular uniform staggered grid where the discrete variables are defined as indicated by Fig. 2.

The discretization of the boundary conditions (2.6) makes use of auxiliary points outside of the domain.

Quantities which are used in gridpoints where they are not defined are computed by linear or bilinear interpolation.

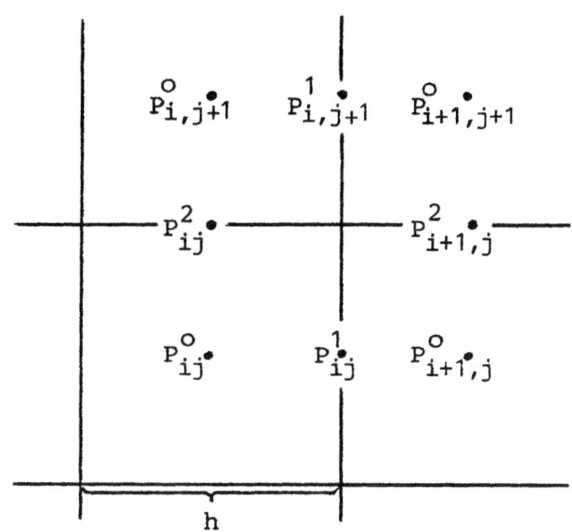

u^h defined at P^1_{ij},

v^h defined at P^2_{ij},

$\rho^h, T^h, p^h, k^h, \varepsilon^h$

defined at P^O_{ij}.

Fig.2: staggered grid

Eqs. (2.1.1-4) and (2.2.1-2) contain the convection-diffusion operator

$$A(c_1, c_2) = c_1 \frac{D}{Dt} - \frac{\partial}{\partial x}(c_2 \frac{\partial}{\partial x}) - \frac{\partial}{\partial y}(c_2 \frac{\partial}{\partial y}) \ . \qquad (3.1)$$

We assume $c_1 > 0$ and $c_2 \geq 0$.

In the following, we discuss the discretization of (3.1) in space and time direction. Besides consistency, stability of the discretization is necessary to obtain convergence of the discrete solution against the continuous solution (at least for linear differential equations). Following [15], we distinguish between *statical* stability which is a property of the space discretization of the stationary equations and *dynamical* statiblity of the time discretization where the space-discretized equations are regarded as a system of ODEs with respect to time.

Spatial discretization

The first derivatives in the convective terms $u\frac{\partial}{\partial x} + v\frac{\partial}{\partial y}$ are approximated by

$$\frac{\partial z}{\partial x}(P^m_{ij}) \longleftarrow \partial^{2h,a}_x z(P^m_{ij}) := \frac{1}{2h} [(1-a)(z(P^m_{i+1,j}) - z(P^m_{ij})) +$$
$$+ (1+a)(z(P^m_{ij}) - z(P^m_{i-1,j}))] \qquad (3.2)$$

and similiarly in the y-direction.

By specifying the value of a in the general expression (3.2) we obtain the standard schemes

$\quad$ a = 0: $\quad$ centered scheme, $\quad$ error $O(h^2)$,

$\quad$ a = 1: $\quad$ backward scheme, $\quad$ error $O(h)$,

$\quad$ a =- 1: $\quad$ forward scheme, $\quad$ error $O(h)$.

In the other terms (e.g. $\frac{\partial p}{\partial x}$) the first derivatives are approximated by centered "short" differences

$$\partial_x^h z(P_{ij}^m) = \partial_x^{h,O} z(P_{ij}^m) = \frac{1}{h}[z(P_{i+0.5,j}^m) - z(P_{i-0.5,j}^m)]$$

At the gridpoints near the boundary, a different discretization is applied for the derivatives in the normal direction (orthogonal to the boundary layer). The formulas are taken from [12] and not given here in detail. If the k-ε model is employed, this is important in order to approximate the generation of turbulence correctly.

Since the second order centered approximation of (3.1) is statically unstable (see below), we consider two algorithmical concepts which are commonly used to guarantee a stable discretization of $A(c_1,c_2)$.

<u>Artifical viscosity</u>

In the concept of (isotropic) artificial viscosity c_2 is replaced by

$$c_2^*(P_{ij}^m) = \max \{c_2, \beta c_1 \max \{|u(P_{ij}^m)|, |v(P_{ij}^m)|\}h\} \text{ and}$$

$$A(c_1,c_2) \longleftarrow A^h(c_1,c_2) = c_1 (\frac{\partial}{\partial t} + u\partial_x^{2h,O} + v\partial_y^{2h,O}) - \tag{3.3}$$

$$- \partial_x^h(c_2^*\partial_x^h) - \partial_y^h(c_2^*\partial_y^h).$$

If $c_2^* > c_2$ the approximation is of order 1 (see also Section 5, Table 1); it can be viewed at as an $O(h)$ disturbance of the continuous problem. The parameter β determines the amount of artificial viscosity introduced and the smoothing properties of the relaxation method [5]. We choose $\beta = 1$; $\beta \geq 0.5$ is necessary for stability.

<u>Upstream-weighted differences</u>

An algorithmically different approach to a statically stable discretization of $A(c_1,c_2)$ is an adaptive upstream-weighted scheme for the convective terms

$$A^h(c_1,c_2) = c_1 (\frac{\partial}{\partial t} + u\partial_x^{2h,a} + v\partial_y^{2h,b}) - \partial_x^h(c_2\partial_x^h) - \partial_y^h(c_2\partial_y^h)$$

where

$$a = \begin{cases} 0 & \text{for } |\xi| \leq 1 \\ 1-1/\xi & \text{for } \xi \geq 1 \\ -1-1/\xi & \text{for } \xi \leq -1 \end{cases}$$

with $\xi = \xi(P_{ij}^m) = uh/c$ if $c := c_2/c_1 \neq 0$; $a = \text{sign}(u)$ if $c = 0$.
b is computed similarly replacing u by v.

In case of $|\xi| > 1$ the upstream-weighted scheme is equivalent
to an anisotropic artificial viscosity with $\beta=(1+|1/\xi|)/2$ in (3.3).

<u>Time discretization</u>
The time derivative is discretized fully implicitly by the
backward Euler method. Other low-order discretization schemes
are not appropriate:
- explicit schemes impose very restrictive limitations on the
  time-step size Δt for (dynamical) stability reasons;
- the unconditionally stable implicit Crank-Nicholson scheme
  has weaker stability properties (no strong absolute stability).
  Our experience indicates that the lack of stability becomes
  apparent if the source in the energy equation is time-
  dependent and no stationary solution of system (2.1) exists.

<u>Stability</u>
To get a little more insight into the stability behaviour of the
difference schemes, we consider the convection-diffusion
operator with constant coefficients
$$A(c_1,c_2) = c_1\left(\frac{\partial}{\partial t} + U\frac{\partial}{\partial x} + V\frac{\partial}{\partial y}\right) - c_2\Delta.$$
A sufficient condition for the statical stability of the
centered difference scheme is the limitation of the *mesh
Reynolds number*

$$R_e^h := Wh/c \leq 4 \qquad (W := |U| + |V|). \qquad (3.4)$$

It is clearly seen, that central differences are (statically)
unstable for the inviscid limit $c = 0$, whatever the size of
Δt and the method of time discretization is. The stabilization
techniques described above increase the viscosity and thereby
fulfill (3.4).

Dynamical stability of the commonly used explizit Euler scheme
is secured if a time-step limitation due to viscosity terms

$$\Delta t \leq h^2/(4c) \tag{3.5}$$

and one due to convection terms

$$\Delta t \leq h/W$$

is met. Together with (3.4) the last condition yields a limitation of Δt not depending on h

$$\Delta t \leq 4c/W^2. \tag{3.6}$$

Condition (3.6) is dominant for $c \longrightarrow 0$, whereas for c large (3.5) determines the effective time-step limitation.

The explicit time discretization of the momentum equations yields an even more restrictive condition of the type

$$\Delta t \leq h/((u^2 + v^2)^{1/2} + a_s),$$

which arises from the pressure term. The ICE-method treats the continuity equation implicitly and thus avoids the speed of sound condition. So the essential limitations for the ICE-methods which is applied in older computer programs like REC-P3 [10] are given by (3.5) and (3.6).

The backward Euler scheme is (dynamically) unconditionally stable. Statical stability requires the boundedness of the inverse of the arising matrix. A sufficient condition is the Z-matrix property which implies (3.4). If centered differences are applied, a weaker *local* stability property which is characterized by h-ellipticity [4] or diagonal dominance of the resulting matrix can be achieved by the time-step limitation

$$\Delta t \leq \begin{cases} h^2/(Wh-4c) & \text{for } Wh-4c > 0 \\ \infty & \text{else} \end{cases} \tag{3.7}$$

The different time-step limitations (3.5) - (3.7) computed with the initial data (2.5) are visualized in Fig. 3.

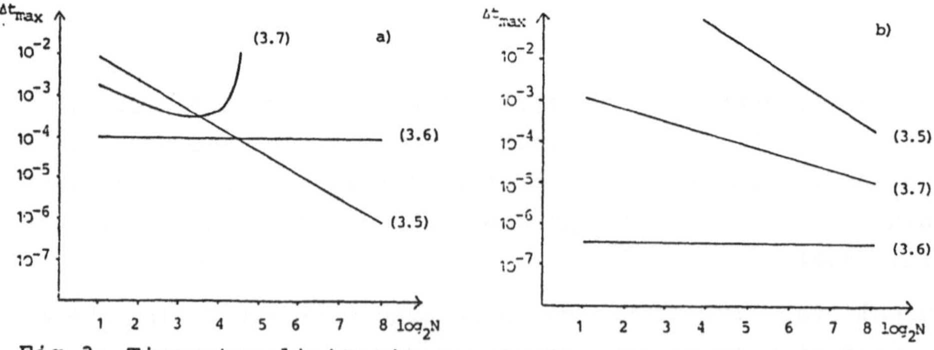

Fig.3: Time-step limitations according to (3.5) - (3.7)
a) $a_\mu = 300$; b) $a_\mu = 1$.

The MG-algorithm described in the next section is based on an unconditionally stable discretization and none of the conditions (3.5) - (3.7) applies.

4. MULTIGRID ALGORITHM

<u>Smoothing</u>

The discrete approximation of system (2.1-2) on the staggered grid can be written as

$$A^h(\rho,\mu)u^h - \frac{1}{3}\partial_x^h(\mu\partial_x^h u^h) - \partial_y^h(\mu\partial_x^h v^h) + \frac{2}{3}\partial_x^h(\mu\partial_y^h v^h) + \tag{4.1.1}$$
$$+ \partial_x^h P^h = 0 \text{ at } P^1_{ij}, \quad (P^h = p^h + \frac{2}{3}\rho^h k^h),$$

$$A^h(\rho,\mu)v^h - \frac{1}{3}\partial_y^h(\mu\partial_y^h v^h) - \partial_x^h(\mu\partial_y^h u^h) + \frac{2}{3}\partial_y^h(\mu\,\partial_x^h u^h) + \tag{4.1.2}$$
$$+ \partial_y^h P^h = 0 \text{ at } P^2_{ij},$$

$$A^h(1,0)\rho^h + \rho^h(\partial_x^h u^h + \partial_y^h v^h) = 0 \text{ at } P^o_{ij}, \tag{4.1.3}$$

$$A^h(\rho c_v, c_p\mu)T^h + P^h(\partial_x^h u^h + \partial_y^h v^h) - \mu B^h(u^h,v^h) = S^h \text{ at } P^o_{ij}, \tag{4.1.4}$$

$$p^h - R\rho^h T^h = 0 \quad \text{at } P^o_{ij}, \tag{4.1.5}$$

$$A^h(\rho,\mu)k^h + \frac{2}{3}\rho^h k^h(\partial_x^h u^h + \partial_y^h v_h) - c_\mu \rho^h(k^h)^2 B^h(u^h,v^h)/\varepsilon^h + \tag{4.1.6}$$
$$+ \rho^h \varepsilon^h = 0 \text{ at } P^o_{ij},$$

$$A^h(\rho,\mu/\sigma_\varepsilon)\varepsilon^h - (1 - \frac{2}{3}c_{\varepsilon_1})\rho^h\varepsilon^h(\partial_x^h u^h + \partial_y^h v^h) - c_{\varepsilon_1}c_\mu\rho^h k^h B^h(u^h,v^h) +$$
$$+ c_{\varepsilon_2}\rho^h(\varepsilon^h)^2/k^h = 0 \text{ at } P^o_{ij}. \tag{4.1.7}$$

The relaxation of (4.1) is a distributive collective scheme which is derived in [5], [6]. One sweep of the relaxation can be summarized briefly as follows

1. Distributive nonlinear relaxation of (4.1.1)

$$u^h \longleftarrow u^h + \hat{u}^h \qquad \text{at } P^1_{ij}$$
$$p^h \longleftarrow p^h + \frac{\mu}{3}\partial_x^h \hat{u}^h \qquad \text{at } P^o_{ij}. \tag{4.2.1}$$

2. Distributive nonlinear relaxation of (4.1.2)

$$v^h \longleftarrow v^h + \hat{v}^h \qquad \text{at } P^2_{ij}$$

$$p^h \longleftarrow p^h + \tfrac{\mu}{3}\partial^h_y \hat{v}^h \quad \text{at } P^0_{ij} \tag{4.2.2}$$

3. Distributive collective nonlinear relaxation of (4.1.3-7)

$$u^h \longleftarrow u^h - \partial^h_x \phi^h \qquad \text{at } P^1_{ij}$$

$$v^h \longleftarrow v^h - \partial^h_y \phi^h \qquad \text{at } P^2_{ij}$$

$$\rho^h \longleftarrow \rho^h + \hat{\rho}^h$$

$$T^h \longleftarrow T^h + \hat{T}^h$$

$$\left.\begin{array}{l} p^h \longleftarrow p^h + \bar{A}^h(\rho,\tfrac{4}{3}\mu)\,\phi^h + \tfrac{2}{3}(\rho^h \hat{k}^h + k^h \hat{\rho}^h) \quad \text{at } P^0_{ij} \\[4pt] k^h \longleftarrow k^h + \hat{k}^h \\[4pt] \varepsilon^h \longleftarrow \varepsilon^h + \hat{\varepsilon}^h \end{array}\right\} \tag{4.2.3}$$

$$\text{with } \bar{A}^h(c_1,c_2) = c_1\left(\tfrac{1}{\Delta t} + u\partial^{2h,a}_x + v\partial^{2h,b}_y\right) - c_2 \Delta^h,$$

$$\Delta^h = (\partial^h_x)^2 + (\partial^h_y)^2 .$$

The motivation for the replacement (4.2.1) becomes clear if we look at equation (4.1.2) and assume μ as locally constant

$$A^h(\rho,\mu)v^h - \tfrac{\mu}{3}(\partial^h_y)^2 v^h - \tfrac{\mu}{3}\partial^h_x(u^h + \hat{u}^h) + \partial^h_y(p^h + \hat{p}^h)$$

$$+ \tfrac{2}{3}\partial^h_y(\rho^h k^h) = 0.$$

If we require that the residual of (4.1.2) is not changed by the change of u, we must change p simultaneously such that

$$-\tfrac{\mu}{3}\partial^h_x \partial^h_y \hat{u}^h + \partial^h_y \hat{p}^h = 0$$

holds. For the changes of p we obtain

$$\hat{p}^h = \tfrac{\mu}{3}\partial^h_x \hat{u}^h .$$

The correction of the pressure guarantees that relaxing (4.1.1) does not disturb the residual of (4.1.2) and vice versa. Similarly, the corrections of u and v in (4.2.3) compensate the change in the residuals of (4.1.1-2) introduced by the replacement of p in (4.2.3). The importance of the distributive relaxation type becomes evident from Table 5 (Section 5).

For the collective relaxation (4.2.3), in each gridpoint a
5x5-system for

$$(\rho^h + \hat{\rho}^h, T^h + \hat{T}^h, \phi^h, k^h + \hat{k}^h, \varepsilon^h + \hat{\varepsilon}^h)$$

which results form the difference operator

$$
\begin{vmatrix}
A^h(1,0) & 0 & -\rho^h \Delta^h & 0 & 0 \\
0 & A^h(\rho c_v, c_p \mu) & -P^h \Delta^h & 0 & 0 \\
-RT^h & -R\rho^h & \bar{A}^h(\rho, \tfrac{4}{3}\mu) & 0 & 0 \\
0 & 0 & -\tfrac{2}{3}\rho^h k^h \Delta^h & A^h(\rho, \mu) & \rho^h \\
0 & 0 & (1-\tfrac{2}{3}c_{\varepsilon_1})\rho^h \varepsilon^h \Delta^h & * & A^h(\rho, \mu/\sigma_\varepsilon)
\end{vmatrix}
$$

with

$$* = -\rho^h(c_{\varepsilon_1} c_\mu B^h(u^h, v^h) + c_{\varepsilon_2}(\varepsilon^h/k^h)^2)$$

is solved.

Each relaxation pass (4.2.1-3) is executed in pointwise Gauss-
Seidel manner with lexicographic (LEX) or red-black (RB) ordering.
Smoothing factors computed by local mode analysis are given in [6].
Line relaxation is not necessary (in the 2D-model) since the flow
is not aligned with the grid lines.

<u>Coarse grid correction</u>
The coarse grid is a staggered grid of mesh-size 2h obtained by
standard coarsening. It is not a subset of the fine grid and
therefore injection in the usual sense is not possible. Residuals
and solutions are restricted to coarser grids using half-weight-
ing operators on the staggered grid as shown in Fig. 4.[7].

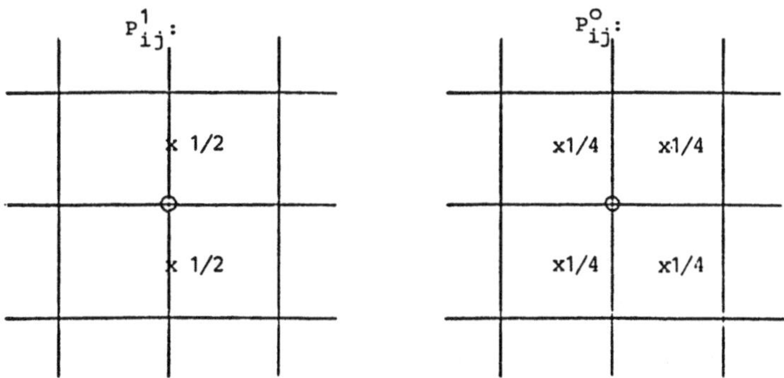

x: fine grid point;   o: coarse grid point

Fig.4 : Restriction operator (e.g. for quantities at P^1_{ij} and P^0_{ij})

The transfer of corrections to finer grids is done by linear interpolation.

The operators on coarser grids result from a stable discretization of the continuous operator on each grid. Usually, they are not restrictions of the fine grid operator (different viscosities or difference stars).

On the coarsest grid (usually 2x2 or 4x4 points), the nonlinear discrete equations are solved by a Newton-like iterative procedure. The matrices of the linear systems are an approximation to the Jacobian and have a sparse but not banded structure. They are efficiently decomposed by the HARWELL-routine MA28. If, however, system (4.1) is highly diagonally dominant (in case of small time-steps) the relaxation procedure described above can also be used to solve the equations on the coarsest grid.

The MG-algorithm MGNS2O which results as a combination of these components is of FAS-type and shows the MG-typical efficiency. The convergence factors for practically meaningful time-step sizes $\Delta t = 10^{-4}$ ($\hat{=}$ 1.8° crank angle at 3000 rpm) are about 0.2 per MG-iteration requiring a computational work of about 4 relaxation sweeps (see Section 5, Table 3). Note that the convergence factors are asymptotically bounded independent of h and Δt, i.e. the algorithm is also an effective solver for the *stationary* compressible Navier-Stokes equations.

NUMERICAL RESULTS

In order to demonstrate the numerical performance of MGNS2O we define three test cases:

Case 1: Describes the gas motion without chemical reaction (S=O in (2.1.4)) using the simple turbulence model (2.4). The interesting quantities are the velocity components, whereas density, temperature and pressure remain nearly constant.

Case 2: The chemical energy is simulated by the source term in the energy equation (2.1.4)

$$S = \begin{cases} 2.26 \cdot 10^{14} \sin(\pi t/3\cdot10^{-3}) & \text{for } 0 \leq t \leq 3\cdot10^{-3} \\ & \text{and } 3 \leq x,\overline{y} \leq 5, \\ 0 & \text{else.} \end{cases}$$

The solution shows large gradients in the temperature field, the maximum temperature exceeding 4000 K. Also ρ and p vary in time and space. μ is given as for Case 1.

Case 3: Similar to Case 1 with the $k - \varepsilon$ model employed and the viscosity given by (2.3).

As mentioned in Section 3, for the convective terms three different space discretizations are used

    1. centered differences,

    2. artificial viscosity,

    3. upstream-weighted scheme.

The size and the order of the corresponding approximation errors are shown in Table 1, which refers to Case 1 with one time-step of size $\Delta t = 10^{-4}$ performed. The table shows the approximation errors in the u-component relative to the coarser grid with double meshsize $\| u^h - u^{2h} \|$ and their ratios. The norm is the discrete L_1-norm, N is the number of grid intervals in each coordinate direction (N=L/h).

$a_\mu = 300:$

N	1		2		3	
8	95.8		66.3		78.7	
		2.0		1.4		1.7
16	47.9		48.8		46.7	
		3.5		1.9		3.8
32	13.7		25.5		12.2	
		4.9		2.0		4.4
64	2.8		12.7		2.8	
		2.9		13.6		2.9
128	1.0		0.9		0.9	

$a_\mu = 1:$

N		2		3	
8		64.5		77.3	
			1.6		1.9
16		40.9		40.9	
			1.8		3.5
32		22.9		11.7	
			1.5		1.9
64		15.5		6.2	
			1.5		1.6
128		10.5		3.8	

Table 1: Error of space discretization

For $N \geq 64$, the different schemes are identical. For smaller N
the centered scheme is of second order accuracy. Although formally unstable at points with large velocity , the lack of
stability does not become evident. Both stable schemes are of
approximation order 1. In case of extremely low viscosity ($a_\mu = 1$),
the stable schemes show the same behaviour as for $a_\mu = 300$,
whereas centered differences are completely unstable.

Moreover, it is clear form Table 1 that for Case 1 a reasonable
approximation with a relative error of about 10^{-3} requires a
resolution of at least 32 x 32 griapoints.

Table 2 shows the analogous approximation errors due to time
discretization on a fixed 32 x 32-grid at the time $t = 512 \cdot 10^{-5}$.
For Case 1, the error of the u-component (E_u), for Case 2 additionally the error of the temperature (E_T) is regarded. Very
large time-steps are not meaningful for Case 2 because they
"jump over" the energy source. In order to achieve a similar
accuracy as in the space discretization, Δt should be about $5 \cdot 10^{-5}$
to 10^{-4}.

$\Delta t \cdot 10^5$	Case 1			Case 2				
	WU	E_u		WU	E_u		E_T	
512	21.3							
256	22.6	44.2	1.6					
128	19.9	28.5	1.7	16.6				
64	17.6	16.6	1.8	13.0	10.0	1.6	75.0	2.9
32	15.6	9.0	1.9	10.5	6.1	1.8	25.8	2.1
16	13.3	4.7	2.0	9.5	3.4	1.9	12.3	2.0
8	10.7	2.4	2.0	8.7	1.7	2.0	6.1	1.9
4	8.2	1.2	2.2	6.9	0.9	2.0	3.2	2.0
2	5.8	0.6	1.7	6.7	0.4	2.0	1.6	2.0
1	5.4	0.3		6.5	0.2		0.8	

Table 2: Error of time discretization (TOL = 10^{-3})

The column "WU" represents the averaged computing work per time-
step of MGNS20, measured in sweeps of the relaxation described in
Section 4. One relaxation sweep is defined as one work unit; the
interpolation and restriction work is not taken into account. The
work per time-step increases slightly as Δt grows (and the diagonal dominance decreases) and approaches an asymptotic value for

$\Delta t \longrightarrow \infty$, i.e. for the solution of the stationary problem. (The MG-iteration stops if the actual residuals in the maximum-norm are on an average less than TOL*(initial residuals).)

	Δt	a_μ	N	V-1/0	V-1/1	V-2/0	V-2/1	W-1/1	RELAX.
LEX-relaxation	10^{-5}	1	32	.38 .48	.16 .50	.18 .53	.09 .55	.14 .61	.897
		300	32	.36 .47	.18 .52	.17 .51	.09 .55	.12 .59	.887
		300	64	.39 .49	.20 .55	.20 .55	.11 .58	.12 .59	.958
	10^{-4}	1	32	.49 .58	.24 .58	.27 .62	.16 .63	.19 .66	.999
		300	32	.43 .53	.25 .60	.25 .59	.18 .65	.17 .64	.997
		300	64	.45 .55	.30 .63	.27 .61	.18 .65	.28 .73	.998
	10^{-3}	300	32	.52 .61	.36 .68	.39 .70	.32 .75	.25 .71	1.
			64	.52 .61	.37 .69	.39 .70	.29 .73	.30 .74	1.
RB-relaxation	10^{-5}	1	32	.36 .47	.13 .47	.18 .53	.08 .53	.10 .57	.897
		300	32	.34 .44	.13 .47	.15 .49	.07 .51	.10 .56	.893
		300	64	.30 .40	.17 .52	.19 .54	.10 .56	.09 .54	.963
	10^{-4}	1	32	.47 .57	.25 .59	.27 .61	.17 .64	.18 .65	.999
		300	32	.44 .54	.27 .61	.27 .61	.19 .66	.17 .65	.998
		300	64	.50 .59	.34 .67	.37 .68	.26 .71	.30 .74	.998
	10^{-3}	300	32	.62 .70	.44 .73	.44 .73	.43 .81	.34 .76	1.
			64	.69 .75	.53 .79	.53 .79	.47 .83	.37 .78	1.

Table 3: Convergence factors of different MG-cycles (max-norm).

More details concerning the convergence properties of different MG-iterations are shown in Table 3. The convergence factors which refer here to Case 2 are observed to be similar in all other examples computed. They are taken as the geometrical average of the reduction factors of the cycles 11-15 or of the last five itera-

tions before reaching round-off level, respectively. The values
in the second line are the convergence factors related to one work
unit (not including grid transfer work). They compare directly to
the convergence factor of the relaxation method when applied as a
solver for system (4.1). "V-K/L" means V-Cycles with K relaxation
sweeps before and L relaxation sweeps after the coarse grid
correction.

For N and Δt large, the relaxation converges extremely slowly,
whereas the convergence factors per work unit of the MG-iteration
are less than 0.7 and not far from the corresponding smoothing
factors (see [6]). The difference between LEX- and RB-ordering
is small. It would be more significant if the characteristics
of the flow were straight lines [6]. Generally, V-cycles seem
to be sufficient and, even for small physical viscosities, the
application of W-cycles does not pay.

In case of a stationary problem with small viscosity, the convergence factors are limited by the quality of the coarse grid
correction. The use of isotropic artificial viscosity implies
a convergence factor ≥ 0.5 [3], the convergence factor with
the upstream-weighted scheme is also bounded away from zero.
In both cases, the coarse grid operator is not identical to
the restriction of the corresponding fine grid operator. Therefore, too expensive MG-cycles are not efficient.

In the following Table 4 we look at the performance of MGNS20
for the Case 3 and the practically important data N=32 or 64
and $\Delta t = 10^{-4}$. Since in a time-dependent problem good starting
values in each time-step are usually available, even a moderate
relative accuracy (TOL=10^{-3}) which is reached in few cycles yields
very good approximations. The convergence of the MG-iteration and
also of the relaxation is observed to be faster in the first
iteration steps and slowly tending to the asymptotic value given
in Table 3. Therefore, for small $\Delta t(\Delta t = 10^{-5}$ to $10^{-4})$ only
one V-1/1-cycle per time-step is needed to fulfill practical
accuracy requirements (relative error of 1%). Although the convergence factors are quite good, the V-1/0-cycle is not competitive.
Its use requires too much grid transfer operations which are not
taken into account in Table 3.

The relative errors given in Table 4 are the differences of the computed solution to an "exact" solution obtained with the same Δt and TOL=10^{-8}, divided by the norm of this "exact" solution. The underlying norm is the max-norm. The CPU-times (in secs) refer to a non-optimized test-version which was run on an IBM 3083-B computer. They are merely a measure for the computational work and should not be compared to CPU-times of other codes.

N	Type of cycle	TOL	relative errors			WU	CPU-time
			u	k	ε		
32	V-1/0	10^{-3}	8.2(-6)	9.9(-6)	1.1(-5)	8.8	286
	V-1/1	10^{-3}	6.4(-6)	8.7(-6)	1.0(-5)	9.5	228
	V-2/1	10^{-3}	6.0(-6)	5.7(-6)	6.7(-6)	12.2	250
	W-1/1	10^{-3}	8.7(-6)	1.8(-5)	1.4(-5)	10.0	244
	V-1/1	1 It.	5.3(-3)	1.2(-2)	8.6(-3)	2.7	69
	V-2/1	1 It.	2.2(-3)	5.1(-3)	3.5(-3)	4.2	87
	RELAX.	10^{-3}				$\approx$ 2000	$\approx$ 8h
64	V-1/1	10^{-3}	1.0(-5)	2.5(-5)	2.1(-5)	10.4	930
	V-2/1	10^{-3}	5.6(-6)	8.7(-6)	6.7(-6)	12.7	982
	V-1/1	1 It.	6.1(-3)	9.1(-3)	8.0(-3)	2.7	261
	V-2/1	1 It.	3.2(-3)	5.6(-3)	4.3(-3)	4.0	330
	RELAX.	10^{-3}				$\approx$ 3500	$\approx$ 60h

Table 4: Results for a full time-course ($5 \cdot 10^{-3}$ secs), $\Delta t = 10^{-4}$, LEX-relaxation, Case 3.

A simpler relaxation method than that used in MGNS2O is a decoupled sweep, first relaxing (4.1.1), then (4.1.2) and then (4.1.3-7) collectively without distributing the corrections.
This corresponds to the relaxation described in Section 4 dropping the pressure corrections in (4.2.1-2) and the velocity corrections in (4.2.3). An MG-iteration based on this relaxation yields convergence factors displayed in Table 5 (Case 2). The importance of a distributive relaxation as a good smoother for system (4.1) becomes apparent.

Δt	distributive	non-distributive
10^{-5}	2	4
10^{-4}	3	23
10^{-3}	5	57

Table 5: Comparison of distributive and non-distributive relaxation. Number of V-2/1-cycles (1 time-step, N=32, TOL=10^{-3})

6. OUTLOOK

As we already pointed out, MGNS2O is an early stage in the development of an MG-solver for the 3D-model. Generalizations and improvements concerning the mathematical model as well as the MG-algorithm are necessary.

1. The following extensions of the mathematical and geometrical model are planned.

i) The inclusion of chemical reactions (combustion) leads to (at least) four additional convection-diffusion equations which express the conservation of chemical species. Also different time scales will occur and must be noticed.

ii) A 3D-model requires one additional momentum equation for the z-velocity component normalized with respect to the piston velocity. The work per gridpoint will slightly increase due to additional terms containing the z-derivatives.

iii) The computational domain is a cylinder with arbitrarily shaped subregions (piston bowl and combustion chamber) attached to the top and bottom end. A boundary fitted grid, i.e. cylindrical coordinates and numerically generated local grids in the subregions, will replace the Cartesian coordinates used in MGNS2O.

2. More sophisticated MG-techniques are under development (line relaxation in z-direction, FAS-FMG, double discretization [6]). Moreover, in the case of 3D-computation with combustion a local grid refinement is necessary in the region of the flame front. In contrast to ADI-methods, adaptive local grid refinements can be done very naturally in the MG context since the FAS-scheme provides information about the relative discretization error.

An effective computation of the 3D-model is limited today by insufficient computer capacity. A reasonable approximation of the gas motion (without combustion) requires a resolution of at least 45 x 45 x 45 ($\approx$ 100.000) gridpoints [9]. (If combustion is included this number will be considerably higher.) Estimates which are based on the performance of MGNS20 indicate that a main memory of about 100 MBytes and a computing speed of several 100 Mflops should be available.

The ultimate aim of the engineers is a numerical method which allows the simulation of series of systematically varied geometrical designs in acceptable computing time. MG-methods provide a powerful numerical tool; together with supercomputers of the next generation numerical simulations will be an important and indispensible aid in engine design.

REFERENCES

[1] Ahmadi-Befrui, B.; Gosman, A.D.; Lockwood, F.C.; Watkins, A.P.: Multidimensional calculation of combustion in an idealised homogeneous charge engine: a Progress Report. no. 810151, Department of Mechanical Engineering, Imperial College of Science and Technology, London, England, 1981.

[2] Becker, K.:
COMFLO - ein Experimentierprogramm zur Mehrgitterbehandlung subsonischer Potentialströmungen um Tragflächenprofile. Preprint no. 604, Sonderforschungsbereich 72, Universität Bonn, 1983.

[3] Börgers, C.:
Mehrgitterverfahren für eine Mehrstellendiskretisierung der Poisson-Gleichung und für eine zweidimensionale singulär gestörte Aufgabe. Diplomarbeit, Institut für Angewandte Mathematik, Universität Bonn, 1981.

[4] Brandt, A.:
Multigrid solvers for non-elliptic and singular-perturbation steady-state problems. Research Report, Dept. of Applied Mathematics, Weizmann Institute of Science, Rehovot, 1981.

[5] Brandt, A.:
Multigrid solutions to steady-state compressible Navier-Stokes equations. I. Computing Methods in Applied Sciences and Engineering V. Proceedings of the Fifth International Symposium, Versailles, France, 14-18 Decembre, 1981 (R. Glowinski, J.L. Lions, eds.), pp. 407-423,1982.

[6] Brandt, A.:
Multigrid Techniques: 1984 Guide with applications to fluid dynamics. Research Report, Dept. of Applied Mathematics, Weizmann Institute of Science, Rehovot, 1984.

[7] Dinar, N.:
Fast methods for the numerical solution of boundary-value problems. PH.D. Thesis, Dept. of Applied Mathematics, Weizmann Institute of Science, Rehovot, 1978.

[8] Gosman, A.,D.:
Prediction of in-cylinder processes in reciprocating
internal combustion engines. Preprint, Department of
Mechanical Engineering, Imperial College of Science and
Technology, London, England, 1983. (Presented at the
6th International Conference on Computing Methods in
Applied Sciences and Engineering, Paris, December 1983)

[9] Gosman, A.D.; Tsui, Y.Y.; Watkins A.P.:
Calculation of three dimensional air motion in model
engines. SAE Technical Paper Series, no. 840229, 1984.

[10] Gupta H.C.; Syed, S.A.:
REC-P3: A computer program for combustion in reciproca-
ting engines. MAE Report no. 1431, Princeton University,
Princeton 1979.

[11] Jameson, A.:
Solution of the Euler equations for two dimensional
transonic flow by a multigrid method. Applied Mathematics and Computation 13, pp. 327-355, 1983.

[12] Launder, B.E.; Spalding, D.B.:
The numerical computation of turbulent flows. Computer
Methods in Applied Mechanics and Engineering 3, pp. 269-
289, 1974.

[13] Linden, J.; Trottenberg, U.:
An MG method for an incompressible Navier-Stokes problem. GAMM-Workshop Efficient Solvers for Elliptic Systems, Kiel, 27-29 January, 1984.

[14] Rice, J.R.; Boisvert, R.F.:
Solving elliptic problems using ELLPACK. Springer-Verlag, 1983.

[15] Roache, P.J.:
Computational fluid dynamics. Hermosa publishers, Albuquerque, N.M. 1972.

[16] Stüben, K.:
MGO1: A multi-grid program to solve $\Delta U - c(x,y)U =
f(x,y)$ (on Ω), $U = g(x,y)$ (on $\partial\Omega$) on nonrectangular
bounded domains Ω. IMA-Report no. 82.02.02, Gesellschaft für Mathematik und Datenverarbeitung, St. Augustin, 1982.

[17] Stüben, K.; Trottenberg, U.:
Multigrid methods: Fundamental algorithms, model problem analysis and applications. Multigrid Methods. Proceedings of the Conference Held at Köln-Porz, November
23-27, 1981 (W. Hackbusch, U. Trottenberg, eds.).
Lecture Notes in Mathematics, 960, pp. 1-176. Springer-
Verlag, Berlin 1982.

THREE-DIMENSIONAL MASS- AND MOMENTUM-CONSISTENT HELMHOLTZ-EQUATION IN TERRAIN-FOLLOWING COORDINATES

Ulrich Schumann and Hans Volkert

DFVLR, Institut für Physik der Atmosphäre
D-8031 Oberpfaffenhofen, Fed. Rep. of Germany

Summary

The derivation of a discrete Helmholtz equation is outlined for the pressure in a fluid dynamic flow model with terrain following coordinates. Special attention is directed towards discrete gradient and divergence operators which are consistent with conservation of mass and momentum. A sketch of a block-iteration scheme for the solution of the three-dimensional pressure equation and of a point- or line-iteration scheme for the two-dimensional boundary problem at the ground follows. Both problems are identified as potentially good applications for multigrid solution methods. Results for a test problem with 65×32×32 grid cells conclude the paper.

Introduction

Airflow over mountains is a topic in mesoscale meteorology that still lacks thorough understanding although important contributions have been made during the last decade. Numerical models have proven to be very powerful tools for studying the flow over mountain barriers as the Rocky Mountains or the Alps, which so far are mostly approximated by a very idealized orography (see e.g. papers by Peltier and Clark [3] or Durran and Klemp [4]; they also contain further references).

The aim is therefore to construct a numerical model which

*   calculates the non-hydrostatic flow over irregular terrain,

*   is deduced and discretized in a general and consistent way from the governing hydrodynamic equations,

- allows for the incorporation of the lower boundary in as correct a way as possible,

- reduces to a simpler model for the case of vanishing topography

- and last, but not least, which uses as efficient numerical schemes as are available.

To approach that goal we start from the unfiltered Navier-Stokes-Equations, use a kind of terrain-following coordinate transformation as proposed by Clark [1], allow for non-equidistant meshsizes in every direction and pay special attention to the discrete gradient and divergence operators concerning consistency with conservation of mass and momentum.

The implicit determination of the pressure field which is consistent with mass continuity results in an elliptic equation. From experience its solution consumes a great proportion of the overall computing time. So very efficient solvers are necessary for that problem.

In the following sections we restrict the discussion to that part of the complete set of equations and of our nomenclature that is necessary to derive the discrete elliptic pressure equation. We shall identify the spatial structure of the discrete Laplace operator in terrain following coordinates, compare it with that of its Cartesian counterpart, pay attention to the boundary condition at the ground and finally we shall outline approaches towards the numerical solution of the discrete pressure equation and present results for a clearly defined test problem.

Coordinate Transformation

Given the non-equidistant meshsizes and the underlaying orography z_s the irregularly bounded physical domain (x',y',z') is mapped onto one of unit cubes $(x,y,z;$ see Fig. 1) by the time-independent transformation:

$$x = x(x')$$
$$(1) \quad y = y(y')$$
$$z = z(\eta) \qquad \eta = z_t(z'-z_s)/(z_t-z_s)$$
$$\text{Note: } \eta(z'=z_s) = 0; \quad \eta(z'=z_t) = z_t \ .$$

In the discrete scheme the monotone functions $x(x')$, $y(y')$ and $z(\eta)$ are defined pointwise. Metric coefficients G^{ij} relate the contra-variant velocity components u^1, u^2, u^3 to their Cartesian counterparts u, v, w (see Appendix A for details):

$$\frac{Dx}{Dt} \equiv u^1 = \frac{\partial x}{\partial x'} u \equiv G^{11} u$$

$$\frac{Dy}{Dt} \equiv u^2 = \frac{\partial y}{\partial y'} v \equiv G^{22} v$$

$$(2) \quad \frac{Dz}{Dt} \equiv u^3 = \frac{\partial z}{\partial x'} u + \frac{\partial z}{\partial y'} v + \frac{\partial z}{\partial z'} w \equiv G^{31} u + G^{32} v + G^{33} w$$

$$\text{with specifically} \quad G^{31} = \frac{\eta - z_t}{z_t - z_s} \frac{\partial z_s}{\partial x'} \frac{\partial z}{\partial \eta}$$

$$G^{32} = \frac{\eta - z_t}{z_t - z_s} \frac{\partial z_s}{\partial y'} \frac{\partial z}{\partial \eta} \quad .$$

The contra-variant components of the velocity vector **V** are particularly useful for the determination of fluxes, as they are defined as the projection of **V** onto the contra-variant base vectors, which are perpendicular to the planes of constant coordinates. The relationship of contra-variant system to Cartesian system (rather than of contra-variant to co-variant system) avoids Christoffel symbols in the transformed momentum equations, which are difficult to handle numerically and which make it at least very complicated to allow for conservation of momentum and kinetic energy.

An important quantity is the Jabobian of the transformation

$$(3) \quad V \equiv \frac{\partial(x',y',z')}{\partial(x,y,z)} = \frac{\partial x'}{\partial x} \frac{\partial y'}{\partial y} \frac{\partial z'}{\partial z} = [G^{11} G^{22} G^{33}]^{-1} \quad .$$

In geometric terms V can be interpreted as a volume-element in the physical system (x',y',z') which is mapped on to a unit cube in the transformed system (x,y,z).

For the transformation of the continuity and momentum equations one needs expressions for the divergence and the gradient operator in the (x,y,z) system. For any scalar ψ one obtains the relations:

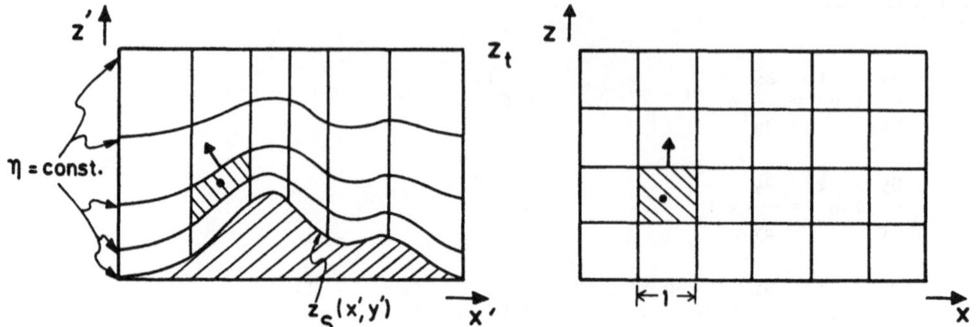

Figure 1. Domain in cartesian (left) and transformed (right) coordinates;
the transformation of one cell with centre (•) and upper normal
velocity component is indicated.

$$(4) \qquad V \, \mathrm{div}(\rho v \psi) = \frac{\partial}{\partial x}(a\psi) + \frac{\partial}{\partial y}(b\psi) + \frac{\partial}{\partial z}(c\psi)$$

with

$r \equiv \rho V$

$v = (u,v,w)$ velocity vector and Cartesian components

and the "density/Jacobian weighted"

contra-variant velocity components

$a \equiv \rho V u^1 = r \, G^{11} \, u$

$b \equiv \rho V u^2 = r \, G^{22} \, v$

$c \equiv \rho V u^3 = r \, G^{31} \, u + r \, G^{32} \, v + r \, G^{33} \, w$

$$(5) \qquad V \, \mathrm{grad}(\psi) = \left(\frac{\partial}{\partial x}[VG^{11}\psi] + \frac{\partial}{\partial z}[VG^{31}\psi], \; \frac{\partial}{\partial y}[VG^{22}\psi] + \frac{\partial}{\partial z}[VG^{32}\psi], \; \frac{\partial}{\partial z}[VG^{33}\psi] \right) \; .$$

Eq. (4) results from the coordinate free definition of the divergence in the
special case of a time-independent coordinate system, while Eq. (5) yields
the Cartesian gradient components and follows from the chain rule for differ-
entiation combined with some "permutation properties" of the metric coefficients:

$$(6) \qquad \frac{\partial}{\partial x}(VG^{11}) = - \frac{\partial}{\partial z}(VG^{31}); \quad \frac{\partial}{\partial y}(VG^{22}) = - \frac{\partial}{\partial z}(VG^{32}); \quad \frac{\partial}{\partial z}(VG^{33}) = 0 \; .$$

We note that the operator in Eq. (4) is in a form which shows clearly that the volume integral $\iiint V\,\mathrm{div}(\rho v\psi)\,dx\,dy\,dz$ can be expressed in terms of surface integrals. This is essential for the conservation of mass ($\psi=1$) and momentum ($\psi=u,v,w$) when difference approximations are introduced (see below). In the same way the integral contribution of the pressure gradient, expressed by Eq. (5) with $\psi=p$, will be momentum conserving. Discretizing the alternative form

$$(5') \quad V\,\mathrm{grad}(\psi) = \left(VG^{11}\frac{\partial\psi}{\partial x} + VG^{31}\frac{\partial\psi}{\partial z}\ ,\ VG^{22}\frac{\partial\psi}{\partial y} + VG^{32}\frac{\partial\psi}{\partial z}\ ,\ VG^{33}\frac{\partial\psi}{\partial z} \right)$$

this would not be so obvious.

We don't give the analytic forms of the transformed equations of motion and continuity, but their discrete analogues, when we have chosen an appropriate grid and defined the discrete operators on it.

Staggered grid and discrete operators

Fig. 2 shows a x-z cross-section of the common staggered grid which allows the approximation of differential quotients with "compact" finite differences of consistency-order 2. Scalars are defined at mesh centres and the velocity components at the centres of the mesh sides in the respective directions. Outside the computational domain a layer of extra meshes is appropriate in every direction for easy specification of boundary values.

A finite difference operator approximates differential quotients; together with an averaging operator it relates quantities between the different classes of points:

$$(7) \quad \delta_n\psi(i,j,k) = \psi(i+I(n),j+J(n),k+K(n)) - \psi(i-I(n),j-J(n),k-K(n))$$

$$\overline{\psi}^{\,n}(i,j,k) = [\psi(i+I(n),j+J(n),k+K(n)) + \psi(i-I(n),j-J(n),k-K(n))] \,/\, 2$$

$$
\begin{aligned}
\text{with} \quad n \ &= \ 1, \quad 2, \quad 3 \\
I(n) &= 0.5, \quad 0, \quad 0 \\
J(n) &= \ \ 0, \ 0.5, \quad 0 \\
K(n) &= \ \ 0, \quad 0, \ 0.5
\end{aligned}
$$

Applying those to the analytic divergence and gradient operators one obtains as difference approximations:

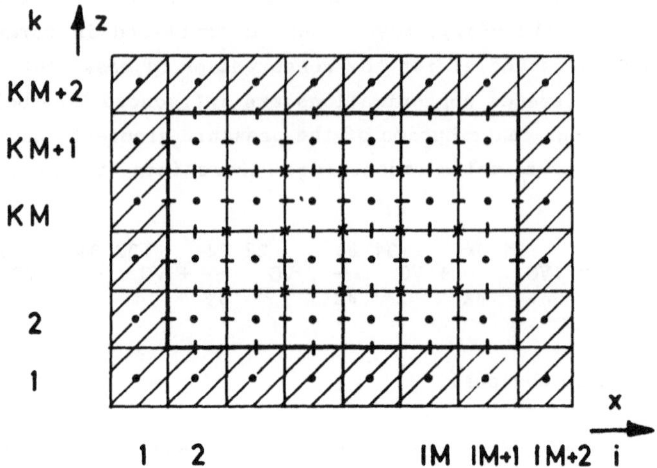

Figure 2. Cross-section of staggered grid in the transformed system (x,y,z)
with computational domain, boundary mesh cells (///),
scalar-points (•), u-points (–), w-points (|) and uw-points (×)

$$(8) \quad V\,\mathrm{GRAD}(p) \;=\; \Big(\delta_1[VG^{11}p]+\delta_3[VG^{31}\overline{\overline{p}}^{\,3}],\; \delta_2[VG^{22}p]+\delta_3[VG^{32}\overline{\overline{p}}^{\,3}],\; \delta_3[VG^{33}p]\Big)$$

$$\equiv\; (\qquad PFU \qquad,\qquad PFV \qquad,\qquad PFW\quad)$$

$$(9) \quad V\,\mathrm{DIV}(\rho v u) \;=\; \delta_1(a\,\overline{u}^{\,1}) \;+\; \delta_2(b\,\overline{u}^{\,2}) \;+\; \delta_3(c\,\overline{u}^{\,3}) \;\equiv\; ADVU$$

$$\text{with}\quad a = G^{11}\,r\,\overline{u}^{\,1}$$

$$(10) \qquad\qquad b = G^{22}\,r\,\overline{v}^{\,2}$$

$$c = G^{31}\,r\,\overline{u}^{\,\overline{31}} \;+\; G^{32}\,r\,\overline{v}^{\,\overline{32}} \;+\; G^{33}\,r\,\overline{w}^{\,3} \;.$$

Expressions for the discrete advection of v and w (ADVV and ADVW) are found
analogous to Eq. (9).

We note the "pure gradient form" of the difference operators above. It
results from the analog formulation of Eqs. (4,5) and implies that sums of
such difference terms over all mesh cells reduce to functions of the boundary

values only. Thus mass and momentum conservation are guaranteed, which means that no unphysical source terms are introduced by the differencing procedure. It can be shown that our discretization of the pressure gradient (Eq.(8)) creates a numerical source term in the budget of kinetic energy. But as the alternative which avoids that (Eq. (5')) doesn't conserve momentum, priority is given to the version above[1] .

Substituting Eqs. (8) and (9) into the equations of continuity and motion (each multiplied by the time-independent Jacobian V) we arrive at the following set of difference equations:

$$(11) \qquad \frac{\partial r}{\partial t} + \delta_1 (a) + \delta_2 (b) + \delta_3 (c) \;=\; 0$$

$$(12a) \qquad \frac{\partial}{\partial t}(r^{-1} u) + ADVU \;=\; - PFU + RU$$

$$(12b) \qquad \frac{\partial}{\partial t}(r^{-2} v) + ADVV \;=\; - PFV + RV$$

$$(12c) \qquad \frac{\partial}{\partial t}(r^{-3} w) + ADVW \;=\; - PFW + RW \quad .$$

The advection terms (ADVU, ADVV, ADVW) and the pressure terms (PFU, PFV, PFW) are defined above, while the remaining terms RU, RV and RW won't be specified here in detail. All three include viscous influences and contributions of the Coriolis force, partly in combination with buoyancy (in RW). The density in $\partial r/\partial t$ is eliminated by $fV(\partial Z(p)/\partial p)(\partial p/\partial t)$ using an equation of state $Z(p)$; f=0 stands for the filtered and f=1 for the unfiltered case. Details are given in Appendix C.

Time discretization is formulated implicitly with respect to the continuity equation and the pressure gradients in the momentum equations. This is necessary for stability reasons. The remaining terms in the momentum equation are treated explicitly with an Euler scheme for the first time level (n=0) and an Adams-Bashforth scheme for all subsequent time levels (n>0; see Appendix B).

---

[1]) For cartesian coordinates a discussion of difference formulae which conserve mass, momentum and kinetic energy can be found in [2].

Discrete pressure equation

A common problem of the integration of fluid dynamic models is the determination of the pressure field p which is in accordance with mass continuity. If we want to calculate p implicitely (with respect to time), we have to substitute expressions for the momenta $\bar{r}^1 u$, $\bar{r}^2 v$ and $\bar{r}^3 w$ at the "new" time level n+1 (these are easily derived from Eqs. 12a-c) into Eq. (11) keeping in mind the relations in Eq. (10). After eliminating the weighted density r with an equation of state r=Z(p) we arrive at a discrete form of Helmholtz's equation, which reads formally (for details see Appendix C)

$$(13) \qquad L_h^{\Omega} \, p_h \;=\; f_h^{\Omega}(x) \qquad\qquad (x \in \Omega) \; .$$

Ω stands for the computational domain (see Fig. 2) and x for any scalar point triplet (i,j,k) in Ω. L_h^{Ω} is a rather complicated 25-point operator (see Fig. 3) due to the "double averages" both in Eq. (8) and in Eq. (10). Its coefficients depend on the meshsizes and the gradients of the orography via the discrete metric coefficients G^{ij} (i,j=1,2,3) and V given in Eqs. (2) and (3). Thus they are not constant in horizontal planes or even the entire domain.

For conservation reasons it is essential to derive L_h^{Ω} from the discrete versions of DIV and GRAD, as outlined above, rather than to discretize an analytically derived Laplace operator for our staggered grid. As a consequence we cannot use the 15-point approximation with compact finite differences for the Helmholtz operator in curvilinear coordinates.

Boundary conditions for the pressure equation

Eq. (13) applies for all grid cells within the computational domain. This system of equations is yet incomplete because it refers to grid point values outside the computational domain. To complete the system, boundary conditions have to be provided. At the sides and at the top this is easily possible for Neumann or Dirichlet boundary conditions as the metric coefficients vanish there[2] . At the bottom the situation is more complicated, as will be shown.

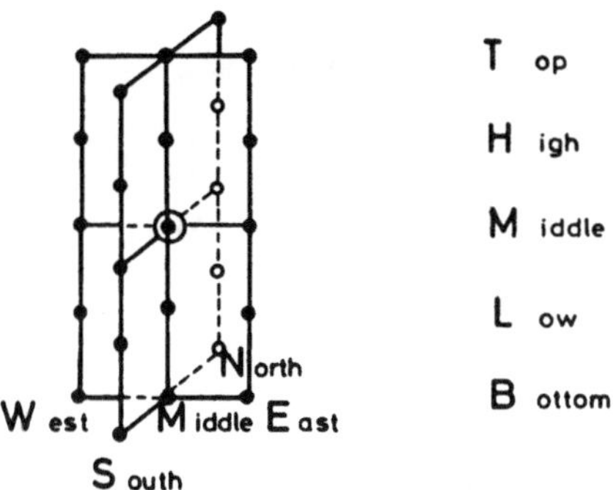

Figure 3.   Orientation of the 25-point operator L_h^{Ω}

Let $\Gamma 1$ be the ground plane $(z\equiv0)$ as part of the entire boundary Γ. On $\Gamma 1$ the following boundary conditions apply:

(14a) $c(z=0) = 0$ and

(14b) $\delta_3^{-1}(\overline{r}\,u)\big|_{z=0} = 0 = \delta_3^{-2}(\overline{r}\,v)\big|_{z=0}$.

Eq. (14a) states that the normal velocity vanishes at the ground. Eq. (14b) is used because it leads to a relatively simple pressure boundary operator $L_h^{\Gamma 1}$ (see [1]).

Substitution of the pressure part of Eqs (12a-c) into Eq. (9) and making use of Eqs. (14a,b) leads to a two-dimensional implicit system involving a diagonally dominant 5-point operator (see Fig. 4); formally that reads:

(15) $L_h^{\Gamma 1}\, p_h = f_h^{\Gamma 1}(x)$ $(x \in \Gamma 1)$.

See Appendix D for further details.

---

[2]) This is true at the top by definition and additionally at the sides, if the orography gradients normal to the boundary vanish; i.e. $z_s(1,j) = z_s(2,j)$ and $z_s(IM+2,j) = z_s(IM+1,j)$ for $j=2,..,JM+1$; $z_s(i,1) = z_s(i,2)$ and $z_s(i,JM+2) = z_s(i,JM+1)$ for $i=2,..,IM+1$.

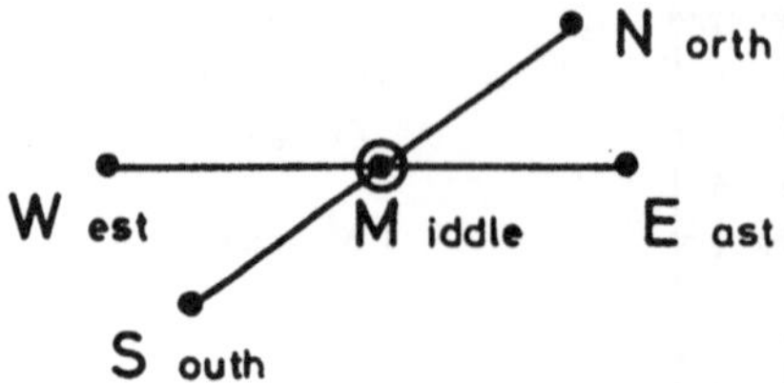

Figure 4. Orientation of the 5-point operator $L_h^{\Gamma 1}$ for the ground level
(k=1)

Approaches towards a solution of the pressure equation

For $z_s \equiv 0$ and $x \equiv x'$, $y \equiv y'$ Eq. (13) reduces to the simpler problem

$$(16) \qquad H_h^\Omega\, p_h \;=\; f_h^\Omega(x) \qquad\qquad (x \in \Omega)\ .$$

where H_h^Ω stands for the standard 7-point operator approximation of the Helmholtz operator with constant coefficients in the x- and y-direction, but with
variable ones in the z-direction. At the ground the standard Neumann boundary
condition is taken into account. For this problem very fast direct solvers
are available based on fast Fourier or cosine transformation in the
y-direction [5] and cyclic reduction combined with Gaussian elimination for
the tridiagonal systems in the x-z planes [6]. For our use we have developed
a subroutine which is fully vectorized for CRAY-computers and takes less than
1 s CPU-time to solve Eq. (16) for a domain with 64^3 grid cells.

Using this solver, a block relaxation scheme is used to compute the solution
of Eqs. (13) and (15) according to the procedure:

$$p^0 \;\equiv\; 0$$

$$(17) \qquad H_h^\Omega\,(p_h^{\nu+1} - p_h^\nu) \;=\; \omega_h^\Omega\,(f_h^\Omega - L_h^\Omega(p_h^\nu)) \qquad\qquad \nu=0,1,..$$

$$H_h^{\Gamma 1}\,(p_h^{\mu+1} - p_h^\mu) \;=\; \omega_h^{\Gamma 1}\,(f_h^{\Gamma 1} - L_h^{\Gamma 1}(p_h^\mu)) \qquad\qquad \mu=0,1,..\ \text{for every}\ \nu\ .$$

Here $H_h^{\Gamma 1}$ represents a point- or a line-relaxation operator which approximates $L_h^{\Gamma 1}$ at the ground. At present we set the relaxation parameters ω^Ω and $\omega^{\Gamma 1}$ to unity. First results for a test problem are presented in the next section.

It is hoped that multigrid methods can be employed alternatively for both the solution of the three-dimensional pressure equation and its two-dimensional boundary problem at the ground. A comparison of the efficiency of both methods in dependency of different underlaying topographies constitutes an interesting research goal.

Test Problem

As a test problem we solve the Poisson equation ($f_p=0$; see Appendix C) with homogeneous Neumann boundary conditions, i.e. the singular problem. It is obvious from Eqs. (A3, A4, C3) that the specific manifestation of the 25-point operator L_h^Ω depends essentially on the underlaying orography and the ratios of typical meshwidths in the three directions. In order to test the relaxation scheme proposed in Eq. (17) we define the following problem, having in mind typical dimensions for mesoscale simulations (in meter):

a) Computational domain:

$$-32500 \leq x' \leq 32500 \qquad \Delta x = \text{const.} = 1000 \qquad \text{IM} = 65 \text{ cells}$$
$$-16000 \leq y' \leq 16000 \qquad \Delta y = \text{const.} = 1000 \qquad \text{JM} = 32 \text{ cells}$$
$$0 \leq z' \leq 10000 \qquad 63 \leq \Delta \eta < 598 \qquad \text{KM} = 32 \text{ cells}$$

$$66560 \text{ cells}$$

The variable vertical meshsize is constructed such that $z(\eta_{k+1/2}) = k-1$, where $z(\eta)$ is given by the relation

$$(18) \qquad z(\eta) \;=\; \text{KM} \left(1 - \frac{\text{artanh}\{(1-\eta/z_t)\sqrt{1-d'}\}}{\text{artanh}\{\sqrt{1-d'}\}}\right)$$

which allows for better resolution in the boundary layer at the ground (see Fig. 6) via the "relative thickness parameter" d (here d = .10; see also [7]).

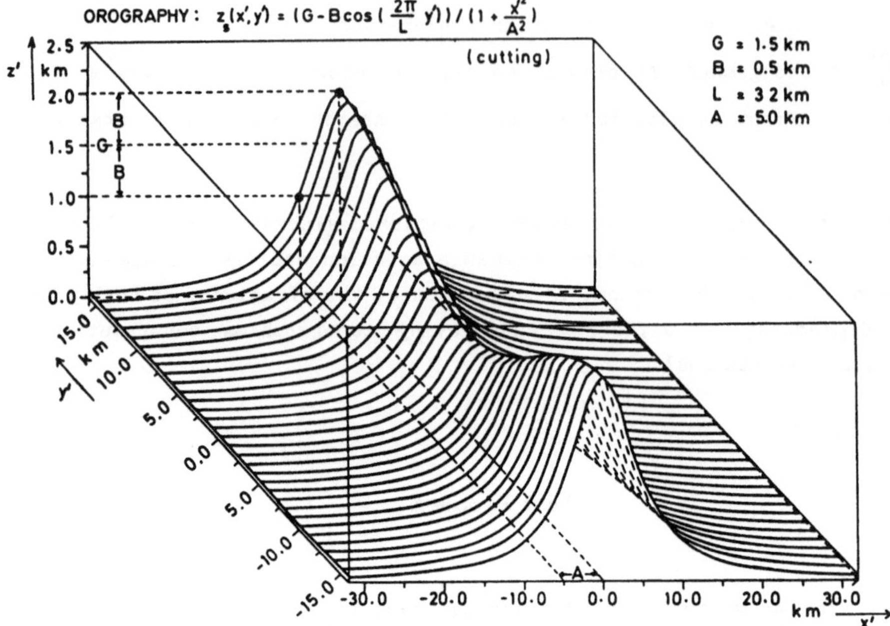

Figure 5. Lower part of physical domain with orography $z_s(x',y')$

b) Orography:

The discrete values for the ground elevation z_s are computed from the formula

$$(19) \quad z_s(x',y') = \frac{G - B\cos(2\pi\, y'/L)}{1 + x'^2 / A^2} \quad \text{with} \quad \begin{aligned} G &= 1500 \text{ m} \\ B &= 500 \text{ m} \\ L &= 32000 \text{ m} \\ A &= 5000 \text{ m} \end{aligned}$$

In the x'-direction z_s exhibits a bell shaped profile, which is determined by the summit height at x'=0 and the half-width A. This type of idealized orography is used in several two-dimensional investigations (e.g. [3,4]). The sinusoidal summit height variation in y'-direction makes the problem fully three-dimensional (see Fig. 5). The two cross-sections in Fig. 6 illustrate size and orientation of mesh cells in the physical domain (note the different scale for every direction).

c) Test procedure:

The test itself consists of five steps

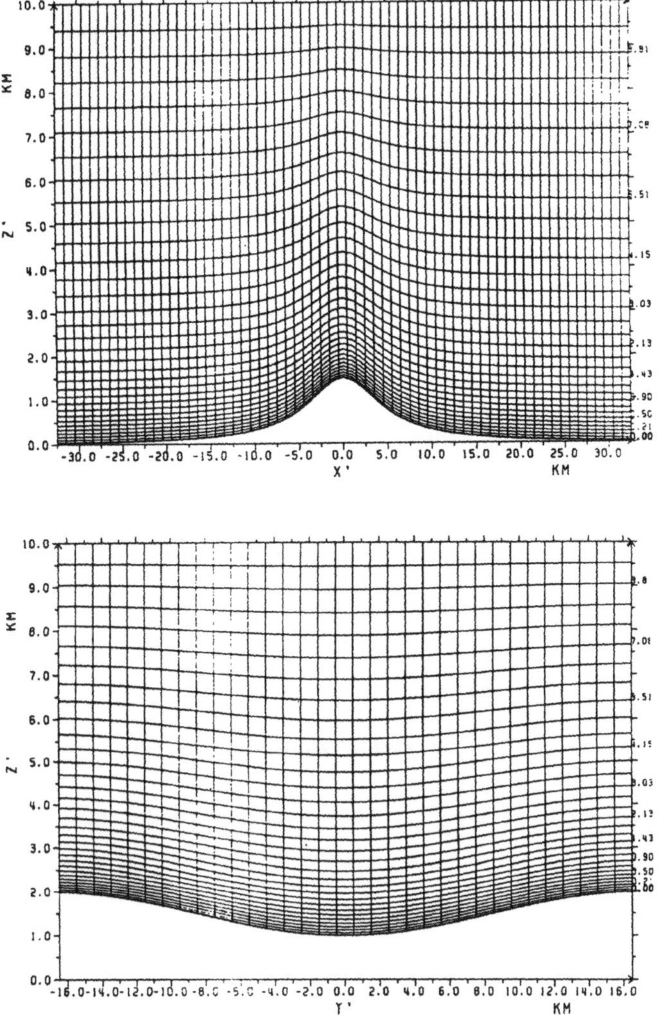

Figure 6. Cross-sections through physical domain at y≡0 (above) and at
x≡±L/4 (below). Mesh cells are bounded by lines x'=const.,
η=const. and y'=const, η=const.

1. prescribe random pressure values p'_{ijk} from the interval $[-\Delta p_{max}, \Delta p_{max}]$
for the computational domain $i=2,..,IM+1;\ j=2,..,JM+1;\ k=2,..,KM+1$

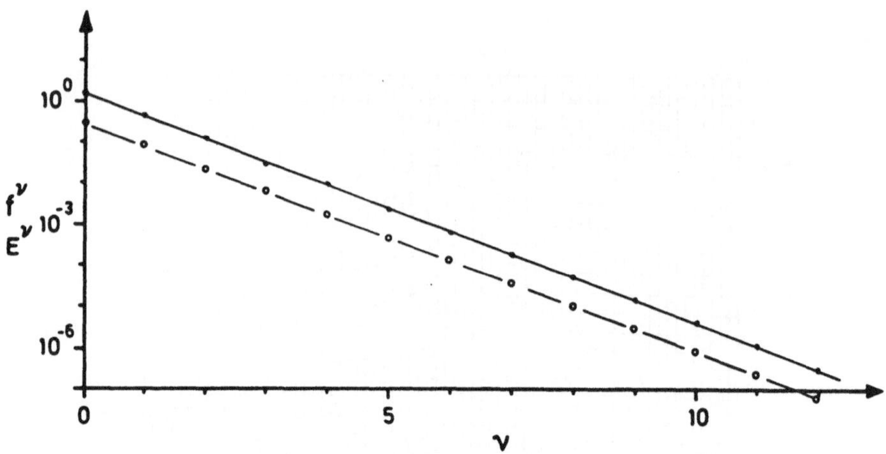

Figure 7.   Relative error measures f^ν ($\bullet$) and E^ν ($\circ$) over iteration number ν

2.   calculate boundary values for i=1,IM+2; j=1,JM+2; k=1,KM+2 according to boundary conditions (Eq. (15) for lower boundary; homogeneous Neumann-type conditions for all other boundaries)

3.   obtain right hand side $f^\Omega_{h\,ijk} = L^\Omega_h\, p'_{ijk}$

4.   calculate p_{ijk} from $f^\Omega_{h\,ijk}$ via the iteration procedure given in Eq. (17)

5.   compare calculated field p_{ijk} with prescribed random field p'_{ijk}

Two similar measures are used to describe the rate of convergence for the iteration process:

$$E^\nu = \max\{|p'_{ijk} - p^{\nu+1}_{ijk}|\} \;/\; \max\{|p'_{ijk}|\}$$

(20)

$$f^\nu = \max\{|p^{\nu+1}_{ijk} - p^\nu_{ijk}|\} \;/\; \Delta p_{max}$$

where the maxima are taken over the entire computational domain. E^ν stands for the maximum relative difference between the prescribed and calculated fields after ν iterations, while f^ν denotes the maximum relative correction between iteration step ν and $\nu+1$.

Fig. 7 shows that both, f^ν and E^ν, are exponentially decreasing functions, as they appear as straight lines when scaled logarithmically. Therefore the error reductions per iteration step

$$(21) \qquad R^\nu_f = f^{\nu+1} / f^\nu \quad ; \qquad\qquad R^\nu_E = E^{\nu+1} / E^\nu$$

are constants. For the test problem we find $R^\nu_f \simeq 0.28 \simeq R^\nu_E$ for $\nu=0,..,12$. We do not report on the computing time yet as the code can still be optimized considerably (concerning vectorization).

The test problem has to be seen as a fairly tough test for the relaxation scheme as the prescribed random pressure field is rougher than meteorological fields, whereas the block-iteration should work best on smooth fields. On the other hand the orography is not very steep. Nevertheless some five iterations per time step will be necessary in mesoscale simulations to obtain a solution sufficiently accurate. Therefore the full vectorization of the algorithm is indispensable.

References

[1] Clark, T.L.:   A small-scale dynamic model using a terrain-following coordinate transformation.
J. Comput. Phys. **24** (1977), pp. 186-214.

[2] Schumann, U.:   Conservation properties of finite difference Euler equations.
GAMM - Wissenschaftliche Jahrestagung, Regensburg, April 1984. ZAMM (1984), to appear.

[3] Peltier, W. and T.L. Clark: Nonlinear mountain waves in two and three spatial dimensions.
Quart. J. R. Met. Soc. **109** (1983), pp. 527-548.

[4] Durran, D. and J. Klemp: A compressible model for the simulation of moist mountain waves.
Mon. Wea. Rev. **111** (1983), pp. 2341-2361.

[5] Wilhelmson, R. and J. Ericksen: Direct solutions for Poisson's equation in three dimensions.

J. Comput. Phys. **25** (1977), pp. 319-331.

[6] Schumann, U. and R. Sweet: A direct method for the solution of Poisson's equation with Neumann boundary conditions on a staggered grid of arbitrary size.
J. Comput. Phys. **20** (1976), pp. 171-182.

[7] Roberts, G.: Computational meshes for boundary layer problems.
in: M. Holt (Ed.), Proceedings of the second international conference on numerical methods in fluid dynamics, Springer Verlag (1971).

Appendices

A: The discrete metric coefficients

The transformation, Eq. (1), and the grid are defined pointwise by input of $x'_{i+1/2} \equiv XH(i)$, $y'_{j+1/2} \equiv YH(j)$ and $\eta_{k+1/2} \equiv HH(k)$ ($i=0,..,IM+1$; $j=0,..,JM+1$; $k=0,..,KM+1$). From these values we compute

$$
\begin{array}{llll}
 & x'_i & \equiv X(i) & = [XH(i) + XH(i-1)] / 2 & i=1,...,IM+1 \\
(A1) & \Delta x'_i & \equiv DX(i) & = XH(i) - XH(i-1) & i=1,..,IM+1 \\
 & \Delta x'_{i+1/2} & \equiv DXH(i) & = X(i+1) - X(i) & i=1,..,IM
\end{array}
$$

and analogous vectors Y, DY, DYH and H, DH, DHH. Further input data are the orographic elevation at cell centres $ZS(i,j) \equiv z_s(x'_i, y'_j)$, and $ZT \equiv z_t$, the elevation of the model top.

The metric coefficients are determined from these parameters. In principle this involves 11 three-dimensional fields, which are defined at different classes of points in the staggered grid (see Fig. 2):

$$
\text{(A2)} \quad
\begin{array}{ll}
G^{11} & \text{u-points} \\[4pt]
G^{22} & \text{v-points} \\[4pt]
G^{33} \\[4pt]
G^{31} & \text{w-points} \\[4pt]
G^{32}
\end{array}
\qquad
\begin{array}{ll}
V \\[4pt]
VG^{11} \\[4pt]
VG^{22} & \text{scalar-points} \\[4pt]
VG^{33} \\[4pt]
VG^{31} & \text{uw-points} \\[4pt]
VG^{32} & \text{vw-points}
\end{array}
$$

Because of storage limitations these fields are reduced to products of two- and one-dimensional quantities (compare Eqs. (2,3)):

$$
\text{(A3)} \quad
\begin{aligned}
G^{11} &= 1./DXH(i) & VG^{11} &= DY(j)*DH(k)*G3(i,j) \\[4pt]
G^{22} &= 1./DYH(j) & VG^{22} &= DX(i)*DH(k)*G3(i,j) \\[4pt]
G^{33} &= (G3(i,j)*DHH(k))^{-1} & VG^{33} &= DX(i)*DY(j) \\[4pt]
G^{31} &= G1(i,j)*(HH(k)-ZT)/DHH(k) & VG^{31} &= (HH(k)-ZT)*VG1(i,j) \\[4pt]
G^{32} &= G2(i,j)*(HH(k)-ZT)/DHH(k) & VG^{32} &= (HH(k)-ZT)*VG2(i,j)
\end{aligned}
$$

$$
V = DX(i)*DY(j)*DH(k)*G3(i,j) \quad .
$$

The 2D-fields are related to the underlaying orography $ZS(i,j)$ as follows:

$$
\text{(A4)} \quad
\begin{aligned}
G1(i,j) &= (ZU(i,j)-ZU(i-1,j)) / (DX(i)*(ZT-ZS(i,j)) \\
&\qquad \text{where} \quad ZU(i,j) = (ZS(i+1,j)+ZS(i,j)) / 2. \\
G2(i,j) &= (ZV(i,j)-ZV(i,j-1)) / (DY(j)*(ZT-ZS(i,j)) \\
&\qquad \text{where} \quad ZV(i,j) = (ZS(i,j+1)+ZS(i,j)) / 2. \\
G3(i,j) &= (ZT-ZS(i,j)) / ZT \\
VG1(i,j) &= (ZS(i+1,j)-ZS(i,j)) * DY(j) / ZT \\
VG2(i,j) &= (ZS(i,j+1)-ZS(i,j)) * DX(i) / ZT \quad .
\end{aligned}
$$

We note that the definitions above are consistent with Eq. (6) if $\partial/\partial x$, $\partial/\partial y$, $\partial/\partial z$ are replaced by δ_1, δ_2, δ_3, respectively. This guarantees that $L_h^{\Omega} \, p_h \equiv 0$ for $p_h \equiv$ const.

B: The time-discretized continuity and momentum equations

The time discretized versions of Eqs. (11,12a-c) read:

$$(B1) \quad \frac{r^{n+1} - r^n}{\Delta t} + \beta\{\delta_1 a^{n+1} + \delta_2 b^{n+1} + \delta_3 c^{n+1}\} + (1-\beta)\{\delta_1 a^n + \delta_2 b^n + \delta_3 c^n\} = 0$$

$$(B2a) \quad (r^{-1} u)^{n+1} = - \gamma \, \Delta t \, \{\delta_1 (VG^{11} p) + \delta_3 (VG^{31} \overline{p^1})\} + R'_u$$

$$(B2b) \quad (r^{-2} v)^{n+1} = - \gamma \, \Delta t \, \{\delta_2 (VG^{22} p) + \delta_3 (VG^{32} \overline{p^2})\} + R'_v$$

$$(B2c) \quad (r^{-3} w)^{n+1} = - \gamma \, \Delta t \, \{\delta_3 (VG^{33} p)\} + R'_w \, .$$

The integration scheme is explicit for $\beta=0$, $\gamma=1$ or $\beta=1$, $\gamma=0$. In these cases the time step is restricted by the speed of sound. The scheme is unconditionally stable with respect to sound waves for $1/2 \leq \beta, \gamma \leq 1$. The fully implicit version $\beta = 1 = \gamma$ is best suited for incompressible flow calculations (filtered version; see Appendix C); in the unfiltered, compressible case sound waves are weakly damped.

The variable p in the equations above stands for the pressure change, $p \equiv p^{n+1} - p^n$. The terms R'_u, R'_v, R'_w contain the momentum sources from the previous time levels, e.g.:

$$(B3) \quad R'_u = (r^{-1} u)^n + \Delta t \, g_1 \, (RU - PFU - ADVU)^n + \Delta t \, g_2 \, (RU - PFU - ADVU)^{n-1}$$

$$n = 0: \quad g_1 = 1 \, , \, g_1 = 0 \qquad \text{(Euler-scheme)}$$
$$n > 0: \quad g_1 = 3/2, \, g_2 = -1/2 \qquad \text{(Adams-Bashforth-scheme)}$$

C: The discrete Helmholtz-equation for the pressure

For the derivation of the discrete pressure equation (13) we need a discretized equation of state ($\rho = Z(p,\theta)$; potential temperature θ; $A_p \equiv \partial Z/\partial p$; $A_\theta \equiv \partial Z/\partial \theta$; filter parameters f_p, $f_\theta \ \varepsilon\ \{0,1\}$) and the transformation for the momenta (see Eq.(4)):

$$(\text{C1}) \quad r^{n+1} - r^n = f_p\, V A_p\, (p^{n+1} - p^n) + f_\theta\, V A_\theta\, (\theta^{n+1} - \theta^n)$$

$$
\begin{aligned}
a^{n+1} &= G^{11}\, (\overline{r}^{\,1} u)^{n+1} \\[4pt]
(\text{C2}) \quad b^{n+1} &= G^{22}\, (\overline{r}^{\,2} v)^{n+1} \\[4pt]
c^{n+1} &= G^{31}\, \overline{(\overline{r}^{\,1} u)^{n+1}}^{\,3} + G^{32}\, \overline{(\overline{r}^{\,2} v)^{n+1}}^{\,3} + G^{33}\, (\overline{r}^{\,3} w)^{n+1}\ .
\end{aligned}
$$

Substituting these and Eqs. (B2a-c) into Eq. (B1) we obtain the following expressions for the Helmholtz operator and for the source term in Eq. (13):

$$
\begin{aligned}
(\text{C3}) \quad L_h^{\Omega}\, p =\ & \delta_1 \{ G^{11} [\delta_1 (VG^{11} p) + \delta_3 (VG^{31}\, \overline{\overline{p}}^{\,3}_1)] \} - (f_p\, V A_p)/(\Delta t^2\, \beta\, \gamma)\, p^n \\[6pt]
& + \delta_2 \{ G^{22} [\delta_2 (VG^{22} p) + \delta_3 (VG^{32}\, \overline{\overline{p}}^{\,3}_2)] \} \\[6pt]
& + \delta_3 \{ G^{31} [\delta_1 (VG^{11} p) + \delta_3 (VG^{31}\, \overline{\overline{\overline{p}}}^{\,3}_1)] + G^{33} \delta_3 (VG^{33} p) \\[6pt]
& \phantom{+\delta_3\{} + G^{32} [\delta_2 (VG^{22} p) + \delta_3 (VG^{32}\, \overline{\overline{\overline{p}}}^{\,3}_2)] \ \}
\end{aligned}
$$

$$
\begin{aligned}
(\text{C4}) \quad f_h^{\Omega}(x) =\ & [\, f_\theta\, V A_\theta\, (\theta^{n+1} - \theta^n) + \Delta t(1-\beta)\{\delta_1 a^n + \delta_2 b^n + \delta_3 c^n\} \,]\ /\ (\beta\gamma\Delta t^2) \\[6pt]
& + [\, \delta_1 (G^{11} R'_u) + \delta_2 (G^{22} R'_v) + \delta_3 (G^{31}\, \overline{\overline{R'}}^{\,3}_u + G^{32}\, \overline{\overline{R'}}^{\,3}_v + G^{33} R'_w) \,]\ /\ (\gamma\Delta t)\ .
\end{aligned}
$$

128

After application of the finite difference and the averaging operators defined in Eq. (7) to Eq. (C4) a rather lenghty expression results for the interrelation between the pressure at the central point 'MM' and the 24 neighbouring points (the subscripts are explained in Fig. 3; e.g. MM $\equiv$ (i,j,k), TE $\equiv$ (i+1,j,k+2)). For the sake of completeness it is given below:

$$L_h^{\Omega}\, p_{MM} = TE\, p_{TE} + TW\, p_{TW} + TN\, p_{TN} + TS\, p_{TS} + (TE+TW+TN+TS)\, p_{TM}$$

$$+ (HE1+HE2)p_{HE} + (HW1+HW2)p_{HW} + (HN1+HN2)p_{HN} + (HS1+HS2)p_{HS}$$

$$+ (HE2+HW2+HN2+HS2+HM)p_{HM}$$

$$\text{(C5)} \qquad + (ME1+ME2)p_{ME} + (MW1+MW2)p_{MW} + (MN1+MN2)p_{MN} + (MS1+MS2)p_{MS}$$

$$+ (ME2+MW2+MN2+MS2+MM)p_{MM}$$

$$+ (LE1+LE2)p_{LE} + (LW1+LW2)p_{LW} + (LN1+LN2)p_{LN} + (LS1+LS2)p_{LS}$$

$$+ (LE2+LW2+LN2+LS2+LM)p_{LM}$$

$$+ BE\, p_{BE} + BW\, p_{BW} + BN\, p_{BN} + BS\, p_{BS} + (BE+BW+BN+BS)\, p_{BM}$$

with the factors:

$$TE = G^{31}_{MM}\, VG^{31}_{HM} / 16; \quad TW = G^{31}_{MM}\, VG^{31}_{HW} / 16$$

$$TN = G^{32}_{MM}\, VG^{32}_{HM} / 16; \quad TS = G^{32}_{MM}\, VG^{32}_{HS} / 16$$

$$HE1 = G^{31}_{MM}\, VG^{11}_{HE} / 4; \quad HE2 = G^{11}_{MM}\, VG^{31}_{MM} / 4 + (G^{31}_{MM}\, VG^{31}_{HM} - G^{31}_{LM}\, VG^{31}_{MM})/16$$

$$HW1 = -\, G^{31}_{MM}\, VG^{11}_{HW} / 4; \quad HW2 = -\, G^{11}_{MW}\, VG^{31}_{MW} / 4 + (G^{31}_{MM}\, VG^{31}_{HW} - G^{31}_{LM}\, VG^{31}_{MW})/16$$

$$HN1 = G^{32}_{MM}\, VG^{22}_{HN} / 4; \quad HN2 = G^{22}_{MM}\, VG^{32}_{MM} / 4 + (G^{32}_{MM}\, VG^{32}_{HM} - G^{32}_{LM}\, VG^{32}_{MM})/16$$

$$HS1 = -\, G^{32}_{MM}\, VG^{22}_{HS} / 4; \quad HS2 = -\, G^{22}_{MS}\, VG^{32}_{MS} / 4 + (G^{32}_{MM}\, VG^{32}_{HS} - G^{32}_{LM}\, VG^{32}_{MS})/16$$

$$HM = G^{33}_{MM}\, VG^{33}_{HM}$$

$$ME1 = VG^{11}_{ME}\,[G^{11}_{MM} + (G^{31}_{MM} - G^{31}_{LM})/4]$$

$$MW1 = VG^{11}_{MW}\,[G^{11}_{MW} - (G^{31}_{MM} - G^{31}_{LM})/4]$$

$$MN1 = VG^{22}_{MN}\,[G^{22}_{MM} + (G^{32}_{MM} - G^{32}_{LM})/4]$$

$$MS1 = VG^{22}_{MS}\,[G^{22}_{MS} - (G^{32}_{MM} - G^{32}_{LM})/4]$$

$$ME2 = G^{11}_{MM}\,(VG^{31}_{MM} - VG^{31}_{LM})/4 - (G^{31}_{MM}VG^{31}_{LM} + G^{31}_{LM}VG^{31}_{MM})/16$$

$$MW2 = - G^{11}_{MW}\,(VG^{31}_{MW} - VG^{31}_{LW})/4 - (G^{31}_{MM}VG^{31}_{LW} + G^{31}_{LM}VG^{31}_{MW})/16$$

$$MN2 = G^{22}_{MM}\,(VG^{32}_{MM} - VG^{32}_{LM})/4 - (G^{32}_{MM}VG^{32}_{LM} + G^{32}_{LM}VG^{32}_{MM})/16$$

$$MS2 = - G^{22}_{MS}\,(VG^{32}_{MS} - VG^{32}_{LS})/4 - (G^{32}_{MM}VG^{32}_{LS} + G^{32}_{LM}VG^{32}_{MS})/16$$

$$MM = -VG^{11}_{MM}\,(G^{11}_{MM}+G^{11}_{MW}) - VG^{22}_{MM}\,(G^{22}_{MM}+G^{22}_{MS}) - VG^{33}_{MM}\,(G^{33}_{MM}+G^{33}_{LM}) - (f_p\,VA_p)/(\beta\gamma\,\Delta t^2)$$

$$LE1 = - G^{31}_{LM}\,VG^{11}_{LE}\,/\,4; \qquad LE2 = - G^{11}_{MM}\,VG^{31}_{LM}\,/\,4 - (G^{31}_{MM}VG^{31}_{LM} - G^{31}_{LM}VG^{31}_{BM})/16$$

$$LW1 = G^{31}_{LM}\,VG^{11}_{LW}\,/\,4; \qquad LW2 = G^{11}_{MW}\,VG^{31}_{LW}\,/\,4 - (G^{31}_{MM}VG^{31}_{LW} - G^{31}_{LM}VG^{31}_{BW})/16$$

$$LN1 = - G^{32}_{LM}\,VG^{22}_{LN}\,/\,4; \qquad LN2 = - G^{22}_{MM}\,VG^{32}_{LM}\,/\,4 - (G^{32}_{MM}VG^{32}_{LM} - G^{32}_{LM}VG^{32}_{BM})/16$$

$$LS1 = G^{32}_{LM}\,VG^{22}_{LS}\,/\,4; \qquad LS2 = G^{22}_{MS}\,VG^{32}_{LS}\,/\,4 - (G^{32}_{MM}VG^{32}_{LS} - G^{32}_{LM}VG^{32}_{BS})/16$$

$$LM = G^{33}_{LM}\, VG^{33}_{LM}$$

$$BE = G^{31}_{LM}\, VG^{31}_{BM} / 16; \quad BW = G^{31}_{LM}\, VG^{31}_{BW} / 16$$

$$BN = G^{32}_{LM}\, VG^{32}_{BM} / 16; \quad BS = G^{32}_{LM}\, VG^{32}_{BS} / 16 \ .$$

Eq. (C5) is valid for k=3,..,KM. For the boundary layer k=2 the condition given in Eq. (14a) has to be taken into account. Formally this means that all terms containing factors G^{3n}_{LM} (n=1,2,3) vanish there. Similarly the G^{3n}_{MM}-terms disappear at level k=KM+1.

D: The discrete pressure boundary equation at the ground

From Eq.(14a,b) and the definition

$$(D1) \quad R'_{w\,1M} = - [\,G^{31}_{1M}(R'_{u\,2M} + R'_{u\,2W}) + G^{32}_{1M}(R'_{v\,2M} + R'_{v\,2W})\,] / (2\,G^{33}_{1M})$$

the following explicit form of the boundary equation for the pressure, Eq. (15), results:

$$(D2) \quad L^{\Gamma 1}_{h}\, p_{1M} = E\, p_{1E} + W\, p_{1W} + N\, p_{1N} + S\, p_{1S} + (E+W+N+S+M)\, p_{1M}$$

with the factors:

$$E = G^{31}_{1M}\, VG^{31}_{1M} \quad ; \quad W = G^{31}_{1M}\, VG^{31}_{1W}$$

$$N = G^{32}_{1M}\, VG^{32}_{1M} \quad ; \quad S = G^{32}_{1M}\, VG^{32}_{1S} \quad ; \quad M = 8\, G^{33}_{1M}\, VG^{33}_{1M}$$

$$
\begin{aligned}
(D3)\quad f_{h\,1M}^{\Gamma_1} ={}& 8\,G_{1M}^{33}\,VG_{2M}^{33}\,p_{2M} \\[4pt]
&+ G_{1M}^{31}\big[4VG_{2E}^{11}p_{2E} - 4VG_{2W}^{11}p_{2W} + VG_{2M}^{31}(p_{3E}+p_{3M}+p_{2E}+p_{2M}) - VG_{1M}^{31}(p_{2E}+p_{2M}) \\
&\qquad\quad + VG_{2W}^{31}(p_{3W}+p_{3M}+p_{2W}+p_{2M}) - VG_{1W}^{31}(p_{2W}+p_{2M})\big] \\[4pt]
&+ G_{1M}^{32}\big[4VG_{2N}^{22}p_{2N} - 4VG_{2S}^{22}p_{2S} + VG_{2M}^{32}(p_{3N}+p_{3M}+p_{2N}+p_{2M}) - VG_{1M}^{32}(p_{2N}+p_{2M}) \\
&\qquad\quad + VG_{2S}^{32}(p_{3S}+p_{3M}+p_{2S}+p_{2M}) - VG_{1S}^{32}(p_{2S}+p_{2M})\big].
\end{aligned}
$$

The numeric subscripts refer to the levels k=1,2,3; the alphabetic ones to the respective horizontal positions, as shown in Fig. 4.

NUMERICAL SOLUTION OF MIXED FINITE ELEMENT PROBLEMS

R. Verfürth

Mathematisches Institut, Ruhr-Universität Bochum,
Postfach 102148, D-4630 Bochum, Germany

SUMMARY

We describe two algorithms for the numerical solution of mixed
finite element problems which arise e.g. from the discretization of the Stokes equations. For the first algorithm we transform the original problem into an equation $Lp = g$ involving a
continuous, positive definite, symmetric linear operator L. We
apply a conjugate gradient algorithm to this equation. The evaluation of Lp is done approximately using a multigrid algorithm.
In the second algorithm we apply the multigrid idea directly
to the indefinite problem. We use Jacobi iteration for the
squared system as smoothing operator. The convergence improves
when using a conjugate residual algorithm. The convergence rate
is measured in a mesh dependent norm. Both algorithms have convergence rates bounded away from 1 independently of the mesh-
size. We present numerical results for the first algorithm.
Numerical experiments for the second algorithm are in progress.

§ 1 INTRODUCTION

We describe two algorithms for the numerical solution of indefinite problems of the form :

Find $(\underline{u}_h, p_h) \in X_h \times M_h$ such that

$$a(\underline{u}_h, \underline{v}_h) + b(\underline{v}_h, p_h) = \ell(\underline{v}_h) \qquad \forall \underline{v}_h \in X_h$$
$$b(\underline{u}_h, q_h) \qquad\qquad = 0 \qquad \forall q_h \in M_h . \qquad (1.1)$$

Here, X_h, M_h are finite dimensional Hilbert spaces, $a : X_h \times X_h$
$\longrightarrow \mathbb{R}$ and $b : X_h \times M_h \longrightarrow \mathbb{R}$ continuous bilinear functionals
and $\ell : X_h \longrightarrow \mathbb{R}$ a continuous linear functional. Moreover, a
has to be X_h-elliptic and b has to satisfy an inf-sup condition
(cf. (2.10) below). Specifically, we are interested in problems

of this type which arise from the discretization of the Stokes
problem

$$- \Delta \underline{u} + \nabla p = \underline{f} \quad \text{in } \Omega , \qquad \underline{u} = 0 \qquad \text{on } \partial\Omega ,$$
$$\text{div } \underline{u} \quad = 0 \quad \text{in } \Omega \tag{1.2}$$

in a simply connected, bounded domain $\Omega \subset \mathrm{IR}^d$, d=2,3.

For the first algorithm we transform (1.1) into an equation
$Lp_h = g_h$ for the 'pressure' which involves a symmetric, positive definite, continuous linear operator $L : M_h \longrightarrow M_h$. We
apply a conjugate gradient algorithm to this problem. Each
evaluation of Lp requires the solution of two discrete Poisson
equations. This is done approximately by applying 2 - 4 iterations of a multigrid algorithm. The resulting iterative process has a convergence rate κ bounded away from 1 independently of the meshsize. Numerical experiments yield values for κ
between .8 and .93.

In the second algorithm we apply the multigrid idea directly
to (1.1). The main difficulty besides the indefiniteness is
the different order of the differntial operators underlying a
and b. This results in a lack of regularity of the pressure.
To overcome this difficulty we introduce a mesh dependent norm.
The convergence rate measured in this norm is bounded away
from 1 independently of the meshsize. The use of a Lanczos algorithm as smoothing operator yields an improvement on Jacobi
relaxation.

§ 2 PRELIMINARIES

Let $H^m(\Omega)$, $m \geq 0$, $H_o^1(\Omega)$ and $L^2(\Omega) := H^o(\Omega)$ be the usual Sobolev
and Lebesgue spaces equipped with the norm

$$\| u \|_m := \{ \int_\Omega \sum_{|\beta| \leq m} | D^\beta u(x) |^2 \, dx \}^{1/2} .$$

We use the same notation for the corresponding product norm on
$H^m(\Omega)^d$. The inner product of $L^2(\Omega)$ is denoted by $(f,g)_o$. For
ease of notation put .

$$X := H_o^1(\Omega)^d \quad , \qquad M := \{ p \in L^2(\Omega) : (p,1)_o = 0 \} \tag{2.1}$$

and introduce the bilinear forms

$$a(\underline{u},\underline{v}) := (\nabla\underline{u},\nabla\underline{v})_0 \quad , \quad b(\underline{v},p) := -(\text{div } \underline{v},p)_0 \qquad (2.2)$$

on XxX and XxM, resp.. Note that

$$|\underline{u}|_1 := a(\underline{u},\underline{u})^{1/2} \qquad (2.3)$$

is a norm on X equivalent to $\|.\|_1$ and that

$$|b(\underline{u},p)| \le \sqrt{d}\ |\underline{u}|_1\|p\|_0 \qquad\qquad \forall\ \underline{u}\in X,\ p\in M\ .\quad (2.4)$$

Moreover, the inf-sup condition

$$\beta := \inf_{p\in M\ \backslash\{0\}} \sup_{\underline{u}\in X\backslash\{0\}} \frac{b(\underline{u},p)}{|\underline{u}|_1\|p\|_0} > 0 \qquad (2.5)$$

holds (cf. Theorem 3.7 in [9]).

Given $\underline{f}\in L^2(\Omega)^d$ the weak form of (1.2) is to find $(\underline{u},p)\in$ XxM such that

$$\begin{aligned} a(\underline{u},\underline{v}) + b(\underline{v},p) &= (\underline{f},\underline{v})_0 &\qquad \forall\ \underline{v}\in X \\ b(\underline{u},q) &= 0 &\qquad \forall\ q\in M\ . \end{aligned} \qquad (2.6)$$

It is well known [9] that (2.6) has a unique solution and that

$$\|\underline{u}\|_2 + \|p\|_1 \le c_0\ \|\underline{f}\|_0 \qquad (2.7)$$

provided $\partial\Omega$ is sufficiently smooth or $\Omega\subset\mathbb{R}^2$ is a convex polygon [5,12].

Let $X_h\subset X$ and $M_h\subset M$, $h>0$, be two families of finite dimensional spaces with $X_{2h}\subset X_h$ and $M_{2h}\subset M_h$, $h>0$, which satisfy the usual approximation assumptions and inverse estimate:

$$\inf_{\underline{v}_h\in X_h} \|\underline{v} - \underline{v}_h\|_\alpha \le c_1\ h^{\beta-\alpha}\|\underline{v}\|_\beta \quad \forall\ \underline{v}\in H^\beta(\Omega)^d,\ 0\le\alpha\le1\le\beta\le2, \quad (2.8a)$$

$$\inf_{p_h\in M_h} \|p - p_h\|_\alpha \le c_2\ h^{\beta-\alpha}\|p\|_\beta \quad \forall\ p\in H^\beta(\Omega)\ ,\ 0\le\alpha\le\beta\le1, \quad (2.8b)$$

$$|\underline{v}_h|_1 \le c_3\ h^{-1}\ \|\underline{v}_h\|_0 \qquad\qquad \forall\ \underline{v}_h\in X_h\ . \qquad (2.9)$$

$c_1,c_2,\ldots$ are generic constants which do not depend on h. The spaces X_h and M_h have to fit together such that

$$\inf_{p_h\in M_h\backslash\{0\}} \sup_{\underline{u}_h\in X_h\backslash\{0\}} \frac{b(\underline{u}_h,p_h)}{|\underline{u}_h|_1\|p_h\|_0} \ge \gamma > 0 \qquad (2.10)$$

holds with a constant γ independent of h. We equip $X_h \times M_h$ with the mesh dependent norm

$$| (\underline{u}_h, p_h) |_h := \{ \| \underline{u}_h \|_o^2 + \| p_h \|_o^2 \, h^2 \}^{1/2} . \tag{2.11}$$

Examples of spaces X_h, M_h satisfying the above assumptions are given in [15] for polygonal domains $\Omega \subset \mathbb{R}^2$. M_h is the space of piecewise linear, continuous functions on a regular triangulation T_h of Ω. X_h is either the space of piecewise linear, continuous functions on $T_{h/2}$ or the space of piecewise quadratic, continuous functions on T_h. Combining the ideas of [4] and [15] these results can be extended to other finite element spaces and regions $\Omega \subset \mathbb{R}^3$.

The approximation of X and M by X_h and M_h then leads to Problem (1.1). Because of (2.3), (2.4) and (2.10) this problem always has a unique solution [9].

§ 3 A POSITIVE DEFINITE PROBLEM FOR THE PRESSURE

To simplify the notation we define the operators

$$B : X_h \longrightarrow M_h \quad , \quad B^\dagger : M_h \longrightarrow X_h \quad ,$$
$$A^{-1} : X_h \longrightarrow X_h \quad , \quad J : L^2(\Omega)^d \longrightarrow X_h$$

as follows

$$(B\underline{u}, p)_o = (\underline{u}, B^\dagger p)_o = b(\underline{u}, p) \qquad \forall \ \underline{u} \in X_h, \ p \in M_h, \tag{3.1}$$
$$a(A^{-1}\underline{u}, \underline{v}) = (\underline{u}, \underline{v})_o \qquad \forall \ \underline{u}, \underline{v} \in X_h, \tag{3.2}$$
$$(J\underline{f}, \underline{v})_o = (\underline{f}, \underline{v})_o \qquad \forall \ \underline{v} \in X_h, \ \underline{f} \in L^2(\Omega)^d. \tag{3.3}$$

Note, that they never need to be computed explicitly. An easy calculation yields

$$\| B\underline{u} \|_o \le \sqrt{d} \ |\underline{u}|_1 \qquad \forall \ \underline{u} \in X_h, \tag{3.4}$$
$$\gamma \ \| p \|_o \le \| B^\dagger p \|_{-1} \le \sqrt{d} \ \| p \|_o \qquad \forall \ p \in M_h, \tag{3.5}$$
$$|A^{-1}\underline{u}|_1 = \| \underline{u} \|_{-1} \qquad \forall \ \underline{u} \in X_h, \tag{3.6}$$

where

$$\|\underline{u}\|_{-1} := \sup_{\underline{v} \in X_h \setminus \{0\}} \frac{(\underline{u},\underline{v})_o}{|\underline{v}|_1} \; .$$

Put

$$L := B\,A^{-1}\,B^{\dagger} \qquad , \qquad \underline{g} := B\,A^{-1}\,J\,\underline{f} \; . \tag{3.7}$$

<u>Lemma 3.1</u>: $L : M_h \longrightarrow M_h$ is a symmetric linear operator satisfying

$$(Lq,q)_o \geq \gamma^2 \|q\|_o^2 \qquad\qquad \forall\, q \in M_h \; , \tag{3.8}$$

$$\|Lq\|_o \leq d\,\|q\|_o \qquad\qquad \forall\, q \in M_h \; . \tag{3.9}$$

The pair $(\underline{u}_h, p_h) \in X_h \times M_h$ is a solution of (1.1) if and only if

$$Lp_h = g \quad \text{and} \quad \underline{u}_h = A^{-1}(J\underline{f} - B^{\dagger}p_h)\,.$$

<u>Proof</u>: The linearity of L is obvious. Equs. (3.4) – (3.6) together with

$$(Lq,r)_o = a(A^{-1}B^{\dagger}q, A^{-1}B^{\dagger}r) \qquad\qquad \forall\, q,r \in M_h$$

imply the symmetry and (3.8),(3.9). The second part of the Lemma follows from the definition of the operators and the unique solvability of (1.1). $\qquad\qquad\qquad\qquad\qquad$ $\square$

The operator L is not known explicitly. Each evaluation of Lp requires the computation of $A^{-1}\underline{w}$ for a suitable $\underline{w} \in X_h$ which is equivalent to the solution of two seperate discrete Poisson equations. This is done approximately by applying n steps of a multigrid (MG-) algorithm with zero starting value to the problem

$$a(\underline{u},\underline{v}) = (\underline{w},\underline{v})_o \qquad\qquad \forall\, \underline{v} \in X_h \; .$$

Denote the resulting approximation to $A^{-1}\underline{w}$ by $K_n\underline{w}$. This defines a linear operator $K_n : X_h \longrightarrow X_h$ with

$$|A^{-1}\underline{w} - K_n\underline{w}|_1 \leq \kappa^n\,|A^{-1}\underline{w}|_1 \qquad\qquad \forall\, \underline{w} \in X_h,\; n \geq 1 \; , \tag{3.10}$$

where κ is the convergence rate of the MG-algorithm.

κ is bounded away from 1 independently of h (cf. e.g. [6,7, 11]). Numerical experiments often yield values $\kappa \leq .1$ (cf. [3, 8,10]). We assume that K_n is symmetric with respect to $(.,.)_o$.

This holds for most MG-algorithms used in practice. Put

$$L_n := B \, K_n \, B^{\dagger} \, . \tag{3.11}$$

Lemma 3.2: Assume that $\kappa^n < \gamma^2 \, d^{-1}$. Then L_n is a symmetric linear operator satisfying for each $q \in M_h$:

$$\| L_n q - Lq \|_o \leq \kappa^n \, d \, \| q \|_o \, , \tag{3.12}$$

$$(L_n q, q)_o \geq (\gamma^2 - d\kappa^n) \, \| q \|_o^2 \, , \tag{3.13}$$

$$\| L_n q \|_o \leq d(1+\kappa^n) \, \| q \|_o \, . \tag{3.14}$$

Proof: The linearity and symmetry of L_n are obvious. Equs.(3.4) -(3.6) and (3.10) imply (3.12). Equs. (3.13), (3.14) immediately follow from (3.12) by using the triangle inequality. $\square$

In the sequel we assume $\kappa^n < \gamma^2 \, d^{-1}$. Since, in principle, γ can be computed explicitly, we could use the above cited bounds for κ to determine the required number n of MG-steps. However, these estimates are far too pessimistic. Our numerical results show that in general 2 - 4 MG-iterations are sufficient.

§ 4 A COMBINED CONJUGATE GRADIENT - MULTIGRID ALGORITHM

The results of §3 show a possible way to solve Problem (1.1) approximately: We compute an approximation $\bar{g}$ to $g = B \, A^{-1} \, J \, \underline{f}$ using the operator K_n and apply an iteration process to the equation $L_n p = \bar{g}$ which only requires the evaluation of $L_n q$. Since L_n is a symmetric, positive definite operator and approximates an operator with the same properties, an appropriate iteration process is the conjugate gradient algorithm.

Algorithm 4.1:
0. *Preprocessing phase* : Compute $\bar{g} := B \, K_n \, J \, \underline{f}$.
1. *Start* : Given an initial guess $p^o \in M_h$ for the pressure p_h solving (1.1). Compute $q^o := L_n p^o$ and put $r^o := q^o - \bar{g}$, $d^o :=$ $-r^o$. Set $i := o$.
2. *Iteration step* : If $\| r^i \|_o \leq \varepsilon$ go to Step 3. Otherwise compute $q^{i+1} := L_n d^i$ and put

$$\alpha^{i+1} := -\frac{(r^i,d^i)_o}{(d^i,q^{i+1})_o} \quad , \quad \beta^{i+1} := \frac{(r^{i+1},r^{i+1})_o}{(r^i,r^i)_o} \quad ,$$

$$p^{i+1} := p^i + \alpha^{i+1} d^i \quad , \quad r^{i+1} := r^i + \alpha^{i+1} \underline{q}^{i+1} \quad ,$$

$$d^{i+1} := -r^{i+1} + \beta^{i+1} d^{i+1} \quad .$$

Replace i by i+1 and return to the beginning of Step 2.

3. *Postprocessing phase* : Compute $\underline{u}^i := K_n(J\underline{f} - B^t p^i)$ and use $(\underline{u}^i,p^i) \in X_h \times M_h$ as final approximation to the solution of (1.1).

From Lemma 3.2 and Exercise 10 of §8.8 in [13] we obtain

$$\|r^i\|_o \leq \frac{2}{\delta}\left(\frac{1-\sqrt{\delta}}{1+\sqrt{\delta}}\right)^i \|r^o\|_o \tag{4.1}$$

where

$$\delta = \frac{\gamma^2 - d\kappa^n}{d(1+\kappa^n)} \leq \frac{\gamma^2}{d} \quad . \tag{4.2}$$

Note, that δ is independent of h. The following error estimate is proved in [16].

<u>Proposition 4.2</u>: Let $(\underline{u}^i,p^i)$ be the last iterate of Algorithm 4.1 and $(\underline{u}_h,p_h)$ be the solution of (1.1). Then we have

$$|\underline{u}_h - \underline{u}^i|_1 + \|p_h - p^i\|_o$$

$$\leq \frac{1}{\gamma^2 - d\kappa^n} \{ 5\varepsilon + 8\kappa^n \|\underline{f}\|_{-1} + 15\kappa^n \|p_h\|_o \} \quad . \tag{4.3} \quad \square$$

The proof of Proposition 4.2 only exploits Lemma 3.1, 3.2 and the stopping criterion $\|r^i\|_o \leq \varepsilon$. Hence it also holds for other iteration processes. The number κ^n in (4.3) is the relative accuracy to which $\bar{q}$, the last residue r^i and the final approximation $\underline{u}^i$ for the velocity are computed. Hence, Steps 1 and 2 of Algorithm 4.1 need only be performed with a moderate accuracy. Once the residue is smaller than ε , we may switch to a higher accuracy in the solution of Poisson's equation. This improves the efficiency of Algorithm 4.1 substantially. Finally, we note that we can use any Poisson solver satisfying (3.10).

§ 5 A MULTILEVEL ALGORITHM

In this section we assume that the regularity assumption (2.7) holds. As usual for multilevel algorithms we have a sequence of meshsizes $h_k = \frac{1}{2} h_{k-1}$, k=1,...,R. Actually we want to solve Problem (1.1) on level R. If no ambiguity can arise, we replace subscripts h_k by k. Instead of (1.1) we have to consider the slightly more general problem

$$a(\underline{u}_k,\underline{v}) + b(\underline{v},p_k) + b(\underline{u}_k,\underline{q})$$

$$= G_k(\underline{v},\underline{q}) \qquad\qquad \forall\ (\underline{v},\underline{q}) \in X_k \times M_k \ . \qquad (5.1)$$

G_k is a linear functional on $X_k \times M_k$ and on the finest grid:

$$G_R(\underline{v},\underline{q}) = (\underline{f},\underline{v})_o \qquad\qquad \forall\ (\underline{v},\underline{q}) \in X_k \times M_k \ . \qquad (5.2)$$

<u>Algorithm 5.1</u>: *(One iteration step at level k, $1 \le k \le R$, with m smoothing steps)*

1. Smoothing : Let $(\underline{u}_k^o, p_k^o) \in X_k \times M_k$ be a given approximation to the solution of Problem (5.1). For $\ell = 1,...,m$ compute the solutions of

$$(\underline{w}_k^\ell, \underline{v})_o + h_k^2\ (r_k^\ell, \underline{q})_o$$

$$= \omega_k^{-2}\ \{G_k(\underline{v},\underline{q}) - a(\underline{u}_k^{\ell-1},\underline{v}) - b(\underline{v},p_k^{\ell-1}) - b(\underline{u}_k^{\ell-1},\underline{q})\}$$

$$\forall\ (\underline{v},\underline{q}) \in X_k \times M_k$$

and

$$(\underline{u}_k^\ell - \underline{u}_k^{\ell-1},\underline{v})_o + h_k^2\ (p_k^\ell - p_k^{\ell-1},\underline{q})_o$$

$$= a(\underline{w}_k^\ell,\underline{v}) + b(\underline{v},r_k^\ell) + b(\underline{w}_k^\ell,\underline{q}) \qquad \forall\ (\underline{v},\underline{q}) \in X_k \times M_k \ .$$

2. Correction : Let $(\bar{\underline{u}}_{k-1}, \bar{p}_{k-1}) \in X_{k-1} \times M_{k-1}$ be the solution of Problem (5.1) with

$$G_{k-1}(\underline{v},\underline{q}) := G_k(\underline{v},\underline{q}) - a(\underline{u}_k^m,\underline{v})$$

$$- b(\underline{v},p_k^m) - b(\underline{u}_k^m,\underline{q}) \quad \forall\ (\underline{v},\underline{q}) \in X_{k-1} \times M_{k-1} \ .$$

If k = 1, put $(\hat{\underline{u}}_{k-1}, \hat{p}_{k-1}) := (\bar{\underline{u}}_{k-1}, \bar{p}_{k-1})$.
If k > 1, compute an approximation $(\hat{\underline{u}}_{k-1}, \hat{p}_{k-1})$ to $(\bar{\underline{u}}_{k-1}, \bar{p}_{k-1})$ by applying μ, $\mu \ge 2$, iterations of the (k-1)-level scheme to (5.1) with starting value zero.

Put

$$\underline{u}_k^{m+1} := \underline{u}_k^m + \hat{\underline{u}}_{k-1} \quad , \quad p_k^{m+1} := p_k^m + \hat{p}_{k-1} \ . \qquad\qquad \Box$$

If we introduce a basis for $X_k \times M_k$, Problem (5.1) can be written in matrix-vector notation as $A_k x_k = d_k$ with a symmetric, indefinite matrix A_k. In the smoothing part of Algorithm 5.1 m Jacobi relaxation steps are applied to the squared system $A_k^2 x_k = A_k d_k$. The relaxation parameter ω_k has to be greater than the spectral radius of A_k.

Let $\delta_{k,m}$ be the convergence rate of one iteration of Algorithm 5.1 at level k with m smoothing steps measured in the $|.|_{h_k}$-norm. Denote by $(\bar{\underline{u}}_k, \bar{p}_k) \in X_k \times M_k$ the solution of Problem (5.1) and by $(\underline{e}_k^\ell, \varepsilon_k^\ell) := (\bar{\underline{u}}_k - \underline{u}_k^\ell, \bar{p}_k - p_k^\ell)$ the error of the ℓ-th iterate. Put $N_k := \dim X_k \times M_k$. There is a complete set of eigenfunctions $(\underline{\phi}_k^j, \psi_k^j)$, $1 \le j \le N_k$, defined by

$$a(\underline{\phi}_k^j, \underline{v}) + b(\underline{v}, \psi_k^j) + b(\underline{\phi}_k^j, q)$$

$$= \lambda_j \{ (\underline{\phi}_k^j, \underline{v})_0 + h_k^2 (\psi_k^j, q)_0 \} \qquad \forall \ (\underline{v}, q) \in X_k \times M_k \ , \qquad (5.4a)$$

$$(\underline{\phi}_k^i, \underline{\phi}_k^j)_0 + h_k^2 (\psi_k^i, \psi_k^j)_0 = \delta_{ij} \qquad 1 \le i, j \le N_k \ . \qquad (5.4b)$$

Because of (2.10) the eigenvalues can be arranged such that

$$0 < |\lambda_1| \le \ \cdots \ \le |\lambda_{N_k}| =: \Lambda_k \ . \qquad\qquad (5.5)$$

An easy calculation [2,17] yields

$$\Lambda_k \le c_4 \ h_k^{-2} \ . \qquad\qquad (5.6)$$

The following analysis holds, if we have $\Lambda_k \le \omega_k$ and $\omega_k = O(h_k^{-2})$. To simplify the notation, we assume that $\Lambda_k = \omega_k$.

We define a scale of mesh dependent norms $\| . \|_s$, $s \in \mathbb{R}$, on $X_k \times M_k$ as follows. Let c_j , $1 \le j \le N_k$, be the coefficients of $(\underline{u}_k, p_k) \in X_k \times M_k$ with respect to the basis $(\underline{\phi}_k^j, \psi_k^j)$, then

$$\| (\underline{u}_k, p_k) \|_s := \{ \sum_{j=1}^{N_k} |\lambda_j|^s c_j^2 \}^{1/2} \ . \qquad\qquad (5.7)$$

Because of (5.4) we have

$$\||(\underline{u}_k, p_k)\||_0 = |(\underline{u}_k, p_k)|_{h_k} . \tag{5.8}$$

It is easy to prove (cf. [2,17]) the *smoothing property*

$$\||(\underline{e}_k^m, \varepsilon_k^m)\||_2 \leq c_5 \; h_k^{-2} \; \frac{1}{\sqrt{2m+1}} \; |(\underline{e}_k^o, \varepsilon_k^o)|_k . \tag{5.9}$$

The crucial point is the *approximation property*

$$|(\underline{e}_k^m - \underline{u}_{k-1}, \varepsilon_k^m - \overline{p}_{k-1})|_k \leq c_6 \; h_k^2 \; \||(\underline{e}_k^m, \varepsilon_k^m)\||_2 . \tag{5.10}$$

For the proof of (5.10) we refer to [17]. Its general structure is similar to that of Bank's convergence analysis [1].
However, his regularity assumptions are not met by the Stokes
problem. Our choice of the norm $|.|_k$ reflects this loss of re-
gularity. Moreover, we have to estimate the velocity and pressure components seperately and to use additional duality arguments.

Using the usual arguments in MG-convergence theory (cf. [2,
7,11,17]), equs. (5.9), (5,10) imply:

<u>Proposition 5.2</u>: The convergence rate of Algorithm 5.1 is boun-
ded by

$$\delta_{k,m} \leq c_7 \; (2m+1)^{-1/2} . \qquad\qquad \forall \; k,m \in \mathbb{N} \tag{5.11} \;\square$$

Instead of Jacobi relaxation we could also perform m steps
of a conjugate residual (CR-) algorithm [14] in the smoothing
part of Algorithm 5.1. Instead of (5.9) we then have [17]:

$$\||(\underline{e}_k^m, \varepsilon_k^m)\||_2 \leq c_8 \; \frac{1}{m+1} \; |(\underline{e}_k^o, \varepsilon_k^o)|_k . \tag{5.12}$$

Since (5.10) is independent of the smoothing procedure, we get
the improved convergence rate

$$\delta_{1,m} \leq \frac{c_9}{m+1} . \tag{5.13}$$

§ 6 NUMERICAL RESULTS FOR ALGORITHM 4.1

We consider three different regions $\Omega \subset \mathbb{R}^2$
   (i) the unit square $\Omega_c := (0,1) \times (0,1)$,

(ii) the L-shaped region $\Omega_L := \Omega_c \setminus (0.5,1) \times (0.5,1)$,
(iii) the slit unit square $\Omega_s := \Omega_c \setminus (0.5,1) \times \{0.5\}$.
and the following right hand sides $\underline{f}$:
example 1 : $\underline{f}^{(1)}(x,y) := \underline{e} := (1,-1)^t$,
example 2 : $\underline{f}^{(2)}(x,y) := 100 \, x(1-x) \, y(1-y) \, \underline{e}$,
example 3 : $\underline{f}^{(3)}(x,y) := 100 \, \exp(-100(x^2+y^2)) \, \underline{e}$.

We use Courant's triangulation with isosceles, rectangular triangles with short sides of length h and put $X_h := S_h^2 \cap X$, $M_h := S_{2h} \cap M$. Here, S_h is the space of continuous, piecewise linear functions on the triangulation with meshsize h.

The Poisson equations are solved with MG-routine HELMH of W. Hackbusch [10]. We choose n between 2 and 4 and $\varepsilon = 10^{-3}$. The MG-iterations terminate, if the L^2-norm of the difference of the last two iterates is smaller than 10^{-7}. According to §4 we compute $\bar{g}$, the last residue r^i and the final approximation $\underline{u}^i$ for the velocity more accurately by replacing n by 4n. In contrast to the theoretical analysis we store the last iterate of the MG-algorithm and use it as starting value for the next call of the MG-routine. This reduces the total number of MG-iterations.

We usually choose $p^0 = 0$ as starting value for Algorithm 4.1. If we have already computed an approximation to p_{2h} on the grid Ω_{2h}, we take its linear interpolant as starting value on the grid Ω_h. We use the meshsizes $h_k = 2^{-k}$, k=2,...,5. All computations were done in single precision arithmetic on the Control Data 175 in Bochum.

Let p^i be the last iterate of Algorithm 4.1. Then

$$\kappa := \{ \| r^i \|_0 \, / \, \| r^0 \|_0 \}^{1/i}$$

is a measure of the mean convergence rate of Algorithm 4.1. In Table 1 we have listed κ for the different examples. A hyphen indicates that $\| r^0 \|_0 \leq \varepsilon$. For further numerical results we refer to [16].

1/h	example 1			example 2			example 3		
	Ω_c	Ω_L	Ω_s	Ω_c	Ω_L	Ω_s	Ω_c	Ω_L	Ω_s
4	.761	.622	.642	.750	.715	.756	–	–	–
8	.507	.635	.583	.800	.801	.636	.841	.763	.815
16	–	.781	.716	.838	.827	.802	.921	.886	.892
32	.709	.816	.840	.842	.890	.800	.920	.833	.859

Table 1 : Mean convergence rate κ of Algorithm 4.1

Acknowledgement: We wish to thank Prof. W. Hackbusch for generously supplying us his programm HELMH.

REFERENCES

1   R.E. Bank: A comparison of two multilevel iterative methods
    for non-symmetric and indefinite elliptic finite element
    equations. SIAM J. Numer. Anal. 18, 724-743 (1981)

2   R.E. Bank, T. Dupont: An optimal order process for solving
    finite element equations. Math. Comp. 36, 35-51 (1981)

3   R.E. Bank, A.H. Sherman: An adaptive multilevel method for
    elliptic boundary value problems. Computing 26, 91-105 (1981)

4   M. Bercovier, O. Pironneau: Error estimates for finite ele-
    ment method solution of the Stokes problem in the primitive
    variables. Numer. Math. 33, 211-224 (1979)

5   C. Bernardi, G. Raugel: Méthodes d'éléments finis mixtes
    pour les équations de Stokes et de Navier-Stokes dans un
    polygone non convexe. Calcolo 18, 255-291 (1981)

6   D. Braess: The convergence rate of a multigrid method with
    Gauss-Seidel relaxation for the Poisson equation. In: Mul-
    tigrid methods (W.Hackbusch, U.Trottenberg; eds.), pp. 368
    -386, Springer 1982

7   D. Braess, W. Hackbusch: A new convergence proof for the
    multigrid method including the V-cycle. SIAM J. Numer. Anal.
    20, 967-975 (1983)

8   A. Brandt: Multi-level adaptive solution to boundary value
    problems. Math. Comp. 31, 333-391 (1977)

9   V. Girault, P.A. Raviart: Finite element approximation of the Navier-Stokes equations. Springer 1979

10 W. Hackbusch: Ein iteratives Verfahren zur schnellen Auflösung elliptischer Randwertprobleme. Report 76-12, Universität Köln 1976

11 W. Hackbusch: Multi-grid convergence theory. In: Multigrid methods (W. Hackbusch, U. Trottenberg; eds.), pp. 177-219, Springer 1982

12 O. Ladyzenskaya: The mathematical theory of viscous incompressible flows. Gordon & Breach 1963

13 D.G. Luenberger: Introduction to linear and nonlinear programming. Addison-Wesley 1973

14 J. Stoer: Solution of large linear systems of equations by conjugate gradient type methods. Preprint

15 R. Verfürth: Error estimates for a mixed finite element approximation of the Stokes equations. RAIRO (to appear)

16 R. Verfürth: A combined conjugate gradient - multigrid algorithm for the numerical solution of the Stokes problem. IMA J. Numer. Math. (to appear)

17 R. Verfürth: A multilevel algorithm for mixed problems. SIAM J. Numer. Math. (to appear)

MULTIGRID SOLUTION OF THE NAVIER-STOKES EQUATIONS IN THE VORTICITY-STREAMFUNCTION FORMULATION

P. Wesseling
Dept. of Mathematics and Informatics,
Delft University of Technology
Postbus 356
2600 AJ Delft
The Netherlands

1. Introduction

The development of fast iterative methods for the vorticity-streamfunction formulation of the Navier-Stokes equations is hampered by difficulties related to the treatment of the boundary conditions. When using the formula of Wood (see Roache (1972)) as a boundary condition for the vorticity no difficulties are encountered using multigrid acceleration of a block-iterative method, which updates vorticity and streamfunction simultaneously. Newton's method is used to handle the nonlinearity, and the multigrid method is employed as an iterative linear systems solver. Numerical experiments on the square cavity flow problem show that the resulting method is considerably more efficient and robust than an earlier method (Roache (1975)) using a fast Poisson solver based on cyclic reduction and fast Fourier transformation, unless the Reynolds number is small.

2. Square cavity flow in vorticity-streamfunction variables

The vorticity-streamfunction formulation of the Navier-Stokes equations in Cartesian tensor notation is given by

$$\psi_{,ii} = \omega , \qquad x_i \in \Omega ,$$

$$u_i \omega_{,i} = Re^{-1} \omega_{,ii} , \quad u_1 = \psi_{,2} , \quad u_2 = -\psi_{,1} , \qquad x_i \in \Omega , \qquad (2.1)$$

$$\psi\big|_{\partial\Omega} = f , \quad \psi_{,n}\big|_{\partial\Omega} = g ,$$

with ψ the streamfunction, ω the vorticity, Re the Reynolds number, u_i the velocity, Ω a two-dimensional domain, and $\psi_{,n}$ the normal derivative of ψ. In the square cavity problem, which will be used here as a test problem, $\Omega = (0,1) \times (0,1)$, $f \equiv 0$, $g = 1$ for $x_2 = 1$, $g = 0$ elsewhere.

3. Finite difference discretization

A uniform computational grid Ω^ℓ is chosen as follows:

$$\Omega^\ell : = \left\{ (x_1,x_2) | x_i = m_i h, \ m_i = 1(1)2^\ell + 1, \ h = (2+2^\ell)^{-1} \right\} . \qquad (3.1)$$

Note that Ω^ℓ lies a distance h within $\partial\Omega$. This is because the boundary conditions will be eliminated from the difference equations. The special choice for h is made in order to facilitate programming of the multigrid method. Coarse grids are obtained by simply doubling the size of the mesh.

Forward and backward difference operators Δ_i and ∇_i are defined by

$$(\Delta_1 \phi)_{ij} : = (\phi_{i+1,j} - \phi_{ij})/h, \quad (\nabla_1 \phi)_{ij} : = (\phi_{ij} - \phi_{i-1,j})/h , \qquad (3.2)$$

and similarly for Δ_2 and ∇_2. A central difference approximation of (2.1) is given by

$$\Delta_i \nabla_i \psi = \bar{\omega}/h^2 , \qquad (3.3)$$

$$\tfrac{1}{2} u_i (\Delta_i + \nabla_i) \bar{\omega} = \mathrm{Re}^{-1} \Delta_i \nabla_i \bar{\omega} , \qquad (3.4)$$

with

$$u_1 = \tfrac{1}{2}(\Delta_2 + \nabla_2)\psi , \quad u_2 = - \tfrac{1}{2}(\Delta_1 + \nabla_1)\psi , \quad \bar{\omega} = h^2 \omega . \qquad (3.5)$$

An upwind difference approximation of (2.1) is obtained if one replaces (3.4) by

$$\tfrac{1}{2} u_i (\Delta_i + \nabla_i) \bar{\omega} = (\gamma_1 \Delta_1 \nabla_1 + \gamma_2 \Delta_2 \nabla_2) \bar{\omega} , \qquad (3.6)$$

with

$$\gamma_i = u_i h \coth(u_i h \mathrm{Re}) . \qquad (3.7)$$

This is the Il'in (1969) scheme (twice rediscovered: Barrett (1974), Chien (1977). It provides a smooth transition, as u_i changes sign, from upwind via central to downwind differences. Unlike standard upwind difference schemes, equation (3.6) has a Fréchet-derivative with respect to u_1 or ψ, which enhances Newton-convergence. The Newton process is applied to the discretized equations.

The boundary conditions are chosen as follows. A difficulty is that we have two boundary conditions for ψ and none for ω. This difficulty is removed by using instead of the Neumann condition for ψ the following formula, due to Woods (1954):

$$\bar{\omega}_w + \tfrac{1}{2}\bar{\omega}_{w+1} - 3\psi_{w+1} = 3hg \ , \qquad \psi_w = 0 \ . \tag{3.8}$$

For a derivation, see Roache (1972), pp. 141, 142. Here w denotes a grid-point on $\partial\Omega$, and w+1 denotes its nearest neighbour in the interior.

The reason for the scaling of ω introduced in (3.5) lies with equation (3.8). Without scaling, the coefficient of ψ_{w+1} would be $-3/h^2$. With the ordering of the unknowns to be introduced shortly this large coefficient appears in the resulting matrix in an off-diagonal position. This is bad for convergence. For complicated problems and methods such as discussed here a rigorous rate of convergence theory is lacking, but it is safe to say that trouble arises when the matrix is far from diagonally dominant. Incidentally, this is of course a prime motivation for using upwind differences. Although the matrix row corresponding to (3.8) is still not diagonally dominant, the unbalance is far less than without scaling of ω. The power of h used in the ω scaling is unique if one requires that (3.3) should not become unbalanced.

4. Newton and Picard iteration

The equations will be solved by nested iteration. Outer iteration handles the nonlinearity, inner iteration (or direct solution) treats the resulting linear systems. In this section only the outer iteration is discussed. Let $\hat{\ }$ indicate values obtained from the preceding iteration. Then the Newton method applied to (3.3), (3.4) and (3.8) results in

$$\Delta_1 \nabla_1 \psi = \bar{\omega}/h^2 \ , \tag{4.1}$$

$$\tfrac{1}{2}\hat{u}_1(\Delta_1+\nabla_1)\bar{\omega} + \tfrac{1}{2}u_1(\Delta_1+\nabla_1)\hat{\bar{\omega}} = Re^{-1}\Delta_1\nabla_1\bar{\omega} + \tfrac{1}{2}\hat{u}_1(\Delta_1+\nabla_1)\bar{\omega} \ , \tag{4.2}$$

148

$$\bar{\omega}_w + \tfrac{1}{2}\bar{\omega}_{w+1} - 3\psi_{w+1} = 3hg \ , \quad \psi_w = 0 \ . \tag{4.3}$$

Newton linearization of the upwind variant (3.3), (3.6) and (3.8), although straightforward, is a little more complicated, because derivatives of γ_i also enter, and we will not write down the resulting formulae.

In each Newton iteration according to (4.1)-(4.3), another linear system has to be solved. The coefficients are variable. In view of the availability of fast standard methods, it is attractive to isolate the Poisson operators in (3.3) and (3.4). This results in Picard iteration:

$$Re^{-1}\Delta_i\nabla_i\bar{\omega} = \tfrac{1}{2}\hat{u}_i(\Delta_i+\nabla_i)\hat{\omega} \ , \tag{4.4}$$

$$\Delta_i\nabla_i\psi = \bar{\omega}/h^2 \ , \tag{4.5}$$

$$\bar{\omega}_w = r(-\tfrac{1}{2}\bar{\omega}_{w+1}+3\psi_{w+1}+3hg) + (1-r)\hat{\omega}_w \ . \tag{4.6}$$

Here r is a relaxation factor, for which a suitable value has to be found by trial and error. This is the LAD method proposed by Roache (1975). For each Picard iteration two Poisson problems have to be solved on a rectangle, which can be done very efficiently.

5. A multigrid method

In order to solve (4.1)-(4.3), or its upwind variant, a multigrid method is used. The basic ingredients of this method are chosen as follows. It is designed to operate as an autonomous subroutine for solving linear algebraic systems. The user provides only the matrix, the right hand side and the termination criterion. Hence, a fixed multigrid schedule is used, for which the sawtooth cycle has been chosen. For a description of this and other multigrid concepts that follow below, see for example Wesseling (1982A,B). Prolongation and restriction are of 9-point type. The coarsest grid always has dimension 3×3. The coarse grid matrices are constructed independently from the user by means of Galerkin approximations. Smoothing is done by means of incomplete factorization (ILU). The last two multigrid ingredients will be described in a little more detail.

The grid points are ordered as in figure 5.1 The unknowns are ordered as

$$
\begin{array}{ccccc}
1+mn-m & . & . & . & mn \\
. & . & . & . & . \\
. & . & . & . & . \\
1+m & 2+m & . & . & 2m \\
1 & 2 & 3 & . & m
\end{array}
$$

<u>Figure 5.1</u> Grid point ordering.

follows: $\psi_1, \psi_2, \psi_3, \ldots, \psi_{mn}, \omega_1, \omega_2, \omega_3, \ldots, \omega_{mn}$. This results in a matrix structure of the following type:

$$
\begin{pmatrix} A_{11} & A_{12} \\ A_{21} & A_{22} \end{pmatrix} \begin{pmatrix} \psi \\ \omega \end{pmatrix} = \begin{pmatrix} b_1 \\ b_2 \end{pmatrix} . \tag{5.1}
$$

The coarse grid Galerkin approximation of a matrix A is given by RAP, with R the restriction and P the prolongation operator. In the case of (5.1) this becomes

$$
RAP = \begin{pmatrix} RA_{11}P & RA_{12}P \\ RA_{21}P & RA_{22}P \end{pmatrix} . \tag{5.2}
$$

Hence, one can employ a subroutine developed for a single elliptic equation to compute RAP. For remarks on programming RAP, see Wesseling (1982A,B).

The sparsity pattern of A_{ij} is conveniently represented by means of the structure of the equivalent finite difference molecule, and is given in figure 5.2.

```
        *                           *  *  *        *  *  *
  *  *  *           *               *  *  *        *  *  *
        *                           *  *  *        *  *  *

        *           *               *  *  *        *  *  *
  *        *     *  *  *            *  *  *        *  *  *
        *           *               *  *  *        *  *  *
```

Finest grid   Coarse grids

<u>Figure 5.2</u> Difference molecules for A_{ij}.

The sparsity pattern for the ILU decomposition to be used for smoothing is on all grids chosen to be equal so that of A on the coarse grids. This is a bit wasteful on the finest grid, where some of the allowed fill-in consists of zero's. The ILU decomposition on the finest grid would not change if the sparsity pattern were to be chosen according to that of figure 5.3, but the resulting saving in work was not taken into account in the operations count

```
*  *
*  *  *        *  *
   *  *           *  *

*  *           *  *
*  *  *        *  *  *
   *  *           *  *
```

<u>Figure 5.3</u> Equivalent ILU sparsity pattern on finest grid.

to be quoted shortly. An algorithm to compute the ILU decomposition is given by (incomplete Crout decomposition):

$$
\begin{aligned}
&\underline{for}\ k\ :\ =\ 1(1)n\ \underline{do}\\
&\underline{begin}\ \underline{for}\ j\ :\ =\ k(1)n\ \underline{while}\ (k,j)\in P\ \underline{do}\\
&\qquad u_{kj}\ :\ =\ a_{kj}\ -\ {\textstyle\sum_1}\,\ell_{kp}u_{pj};\\
&\qquad \underline{for}\ i\ :\ =\ k+1(1)n\ \underline{while}\ (i,k)\in P\ \underline{do}\\
&\qquad \ell_{ik}\ :\ =\ (a_{ik}\ -\ {\textstyle\sum_2}\,\ell_{ip}u_{pk})/u_{kk}\\
&\underline{end};
\end{aligned}
$$

Here P is the chosen sparsity pattern, and the range of the sums $\sum_1$ and $\sum_2$ is given by

$$
\sum_1 :\ p\in\{1,2,\ldots,k-1\}\ \wedge(k,p)\in P\ \wedge\ (p,j)\in P\ ,
$$

$$
\sum_2 :\ p\in\{1,2,\ldots,k-1\}\ \wedge(i,p)\in P\ \wedge\ (p,k)\in P.
$$

The elements of the incomplete triangular factors L and U are given by ℓ_{ij} and u_{ij}. With the foregoing algorithm, diag(L) = I.

Let the linear system to be solved be denoted symbolically by

$$
Au = b\ . \tag{5.3}
$$

One application of ILU-smoothing is given by

$$
u\ :\ =\ u + (LU)^{-1}(b-Au)\ . \tag{5.4}
$$

Clearly, ψ and ω are updated simultaneously in this process.

One might ask what other, perhaps simpler, smoothing processes would be applicable. For the present problem, the smoothing process should satisfy the following requirements:

-It should work for convection-diffusion problems. From the catalogue of smoothing factors compiled by Kettler (1982) one can select possible candidates. Probably one could equally well use line-Gauss-Seidel relaxation, with lines and marching directions in all possible combinations.

-Simultaneous updating of ψ and ω.

6. Numerical experiments and conclusions

For every computation we started with the zero solution. The initial iterand for the inner iteration was the result of the preceding outer iteration. The inner iterations were terminated when, in the notation of equation (5.3),

$$||Au^n-b|| < 10^{-6}||b|| , \qquad (6.1)$$

u^n being the result of the nth (inner) iteration and $||.||$ the maximum norm. Of course, for the sake of efficiency, in a production code one would choose the desired precision of the inner iterations variable and dependent on the precision reached in the outer iteration under execution. The rate of convergence was measured by the quantity

$$\rho : = \{||au^n-b||/||Au^0-b||\}^{1/n} . \qquad (6.2)$$

The rate of convergence of the Newton iterations was monitored by means of $||u^{n+1}-u^n||$, u^n now being the result of the nth Newton iteration. Iteration was terminated when $||u^{n+1}-u^n|| < 10^{-5}$. The Newton method always converged quadratically. The Picard iterations were terminated when $||u^{n+1}-u^n|| < 10^{-6}||u^1-u^0||$. Because of the observed rather slow linear convergence, the resulting accuracy is certainly not better than for the Newton method.

The computational cost of the two methods considered is estimated by counting floating point operations (flops) in the mathematical formulae defining the algorithms. This enables one to make a rough guess of the relative amount of CPU time. We assume that for the Poisson equations in the Picard method a fast Poisson solver of cyclic reduction-fast Fourier transformation is used. There are different variants of this type of method. Various possibilities are reviewed by Dorr (1970), from which publication we deduce that a typical operations count would be $10\ell(2^{\ell}+1)^2$ flops, for solving one Poisson problem. In addition evaluation of the right hand sides take $7(2^{\ell}+1)^2$ flops, so that we obtain as estimate for the computational cost of one Picard iteration $(7+20\ell)(2^{\ell}+1)^2$ flops.

The computational cost of the multigrid method used is about $259(2^{\ell}+1)^2$ flops per iteration, with a preliminary work of $728(2^{\ell}+1)^2$, needed to construct the coarse grid matrices and the incomplete factorizations. For further details, see Wesseling and Sonneveld (1980).

Table 6.1 contains results concerning the rate of convergence and computational cost of the nested Newton iteration–multigrid iteration method, whereas table 6.2 contains results for the Picard iteration – fast Poisson solver method. For further details, see Wesseling and Sonneveld (1980) and Mol (1981). In table 6.1, n is the number of the Newton iteration, and a, b, c stands for a = number of multigrid iterations carried out, b = ρ as defined by (6.2), c = total number of flops (preliminary + iteration work). In table 6.2, a, b, c stands for a = r(the under–relaxation factor in (4.6)), b = number of Picard iterations carried out, c = total number of flops. By a'4 we mean a $\times$ 10^4.

Re	n/ℓ	2	3	4	5
10	1	5, .039, 4.7'4	5, .047, 1.7'5	5, .044, 6.3'5	5, .043, 2.4'6
	2	4, .032, 4.1'4	4, .038, 1.5'5	4, .044, 5.5'5	4, .045, 2.1'6
	3	2, .036, 3.0'4	2, .051, 1.1'5	1, .041, 3.2'5	1, .11, 1.2'6
	4	1, .051, 2.5'4	–	–	–
50	1	5, .039, 4.7'4	5, .047, 1.7'5	5, .044, 6.3'5	5, .043, 2.4'6
	2	6, .062, 5.3'4	5, .054, 1.7'5	5, .053, 6.3'5	4, .044, 2.1'6
	3	5, .065, 4.7'4	4, .075, 1.5'5	3, .051, 4.7'5	3, .067, 1.8'6
	4	2, .11, 3.0'4	1, .079, 0.9'5	1, .12, 3.2'5	1, .084, 1.2'6
100	1	5, .039, 4.7'4	5, 0.47, 1.7'5	5, .044, 6.3'5	5, .043, 2.4'6
	2	7, .12, 5.8'4	7, .13, 2.2'5	5, .079, 6.3'5	5, .072, 2.4'6
	3	7, .16, 5.8'4	5, .088, 1.7'5	3, .059, 4.7'5	4, .068, 2.1'6
	4	5, .15, 4.7'4	3, .099, 1.3'5	1, .078, 3.2'5	1, .085, 1.2'6
	5	2, .23, 3.0'4	1, .11, 0.9'5	–	–
150	1	7, .099, 5.8'4	6, .083, 1.9'5	6, .080, 7.1'5	6, .081, 2.7'6
	2	6, .092, 5.3'4	5, .084, 1.7'5	5, .083, 6.3'5	5, .082, 2.4'6
	3	3, .079, 3.0'4	3, .078, 1.3'5	2, .064, 4.0'5	2, .063, 1.5'6

Table 6.1 Results for nested Newton iteration–multigrid iteration method.

Re/ℓ	2	3	4
10	.276, 21, 2.5'4	.175, 26, 1.4'5	.092, 36, 7.8'5
20	–	.170, 27, 1.4'5	–
40	–	.089, 54, 2.4'5	–

Table 6.2 Results for Picard-fast Poisson solver method.

In table 6.1, the results for Re=10, 50, 100 were obtained with central differences. For Re=150 no convergence was obtained, and upwind differences were used. For Re=150 the termination criteria for the Newton iterations and the multigrid iterations were slightly different from those for the lower Reynolds numbers, cf. Mol (1981). As noted before, the Newton iterations converged quadratically. Clearly, the rate of convergence of the multigrid method is insensitive to the switch from central to upwind differences, and to the changes in the coefficients induced by the Newton method. Furthermore, the computational complexity is proportional to the number of finest grid-points, in accordance with multigrid theory. We expect that with upwind differences Re could be increased arbitrarily; of course, the "numerical Reynolds number" would remain the same.

In table 6.2, the value of the under-relaxation factor r is the best value obtained from a large number of experiments. Where no entry is given, no convergence could be obtained. No convergence was obtained for Re=50 and Re=100. As is to be expected, the Picard method deteriorates as Re increases. For Re small enough the method is cheaper than the Newton-multigrid method. However, the cost is very sensitive to the value of r, the best value of which is unknown a priori.

References

Barrett, K.E., The numerical solution of singular-perturbation boundary-value problems. J. Mech. Appl. Math. 27, 57-68, 1974.

Chien, J.C., A general finite difference formulation with application to the Navier-Stokes equations. Comp. Fl. 5, 15-31, 1977.

Dorr, F.W., The direct solution of the discrete Poisson equation on a rectangle. SIAM Review 12, 248-263, 1970.

Il'in, A.M., Differencing scheme for a differential equation with a small parameter affecting the highest derivative. Math. Notes Acad. Sc. USSR 6, 596-602, 1969.

Kettler, R., Analysis and comparison of relaxation schemes in robust
multigrid and preconditioned conjugate gradient methods. In: W. Hackbusch,
U. Trottenberg (eds.), Multigrid methods. Proceedings, Köln-Porz, Nov.
1981. Lecture Notes in Math. 960, 502-534, Springer-Verlag, Berlin etc.
1982.

Mol, W.J.A., Numerical solution of the Navier-Stokes equations by means of
a multigrid method and Newton-iteration. In: W.C. Reynolds, R.W. MacCormack
(eds.): Seventh Int. Conf. on Numerical Methods in Fluid Dyn. Proceedings,
Stanford 1980. Lecture Notes in Physics 141, 285-291. Springer-Verlag,
Berlin etc. 1981.

Roache, P.J., Computational Fluid Dynamics. Hermosa Publishers,
Albuquerque, 1972.

Roache, P.J., The LAD, NOS and Split NOS methods for the steady-state
Navier-Stokes equations. Computers and Fluids 3, 179-196, 1975.

Wesseling, P. and Sonneveld P., Numerical experiments with a multiple grid
and a preconditioned Lanczos type method. In: R.Rautmann (ed.),
Approximation methods for Navier-Stokes problems. Proceedings, Paderborn
1979. Lecture Notes in Mathematics 771, 543-562. Berlin etc., Springer-
Verlag 1980.

Wesseling, P., A robust and efficient multigrid method. In: W. Hackbusch,
U. Trottenberg (eds.), Multigrid methods. Proceedings, Köln-Porz Nov. 1982.
Lecture Notes in Math. 960, 614-630. Springer-Verlag, Berlin etc. 1982A.

Wesseling, P., Theoretical and practical aspects of a multigrid method.
SIAM J. Sci. Stat. Comp. 3, 387-407, 1982B.

Woods, L.C., A note on the numerical solution of fourth order differential
equations. Aeronautical Quarterly 5, Part 3, 176, 1954.